# Sensory Shelf Life Estimation of Food Products

# Sensory Shelf Life Estimation of Food Products

## Guillermo Hough

**CRC Press**
Taylor & Francis Group
Boca Raton  London  New York

CRC Press is an imprint of the
Taylor & Francis Group, an **informa** business

CRC Press
Taylor & Francis Group
6000 Broken Sound Parkway NW, Suite 300
Boca Raton, FL 33487-2742

© 2010 by Taylor and Francis Group, LLC
CRC Press is an imprint of Taylor & Francis Group, an Informa business

Printed in the United States of America on acid-free paper
10 9 8 7 6 5 4 3 2 1

International Standard Book Number: 978-1-4200-9291-2 (Hardback)

**Library of Congress Cataloging-in-Publication Data**

Hough, Guillermo.
Sensory shelf life estimation of food products / author, Guillermo Hough.
p. cm.
"A CRC title."
Includes bibliographical references and index.
ISBN 978-1-4200-9291-2 (hardcover : alk. paper)
1. Food--Sensory evaluation. 2. Food--Shelf-life dating. 3. Detectors. 4. Food spoilage--Risk assessment. 5. Food--Testing. 6. Food industry and trade--Quality control. I. Title.

TX546.H68 2010
664'.072--dc22                                                          2010011677

**Visit the Taylor & Francis Web site at**
**http://www.taylorandfrancis.com**

**and the CRC Press Web site at**
**http://www.crcpress.com**

# DEDICATION

*To my parents Mary and Ken,*
*To my wife Adriana,*
*To my children Julieta, Natalia and Tomás.*

# Contents

# *Preface*

In the year 2000 we started a four-year project on sensory shelf life (SSL) financed by Ciencia y Tecnología para el Desarrollo (CYTED), Madrid, Spain. The project included participants from Argentina, Chile, Colombia, Costa Rica, Spain, and Uruguay. Thanks to this project we were able to develop most of the methodology that is presented in this book. One of the most valuable collaborations came from a nonfood-science group from Cataluña in Spain directed by Dr. Guadalupe Gómez. This group specializes in survival analysis statistics and at the time worked with data from AIDS infection and had no connection with SSL methodology. The exchange with this group allowed the application of survival analysis methodology applied to SSL, which comprises more than three chapters of this book. As a result of the CYTED project, a book was published (Hough, G. and S. Fiszman, ed. 2005. *Estimación de la Vida Util Sensorial de los Alimentos*. Madrid: programa CYTED) which was a worthy predecessor of the present book.

This book starts with an introduction (Chapters 1 and 2) which includes definitions and a review of books published on the theme of food shelf life and the basics of sensory analysis and how they apply to shelf life studies. Experimental design aspects are covered in Chapter 3. Survival analysis methodology is covered extensively with the basic model (Chapter 4) and its extensions (Chapter 5). Detailed instructions and software functions are presented which will allow readers to perform their own SSL estimations. Data sets used in examples and the R statistical package functions can be downloaded from the editor's website (http://www.crcpress.com/). The cut-off point methodology used to estimate SSL is presented in Chapter 6. Many researchers have a special interest in accelerated shelf-life testing and the methodology is covered in Chapter 7 including R function to perform a non-linear regression to better estimate activation energy. Potential pitfalls of accelerated studies are discussed. Finally, Chapter 8 presents extensions of survival analysis statistics to other areas of food quality. Optimum concentrations of ingredients and optimum cooking temperatures are among examples presented.

# Acknowledgments

Thanks to Susana Fiszman for sharing the editing of a book that served as foundation to this one; Guadalupe Gómez, Malu Calle, Klaus Langohr, and Carles Serrat for believing survival analysis could be applied to SSL; and Lorena Garitta, Mercedes López-Osornio, and Miriam Sosa, present students and collaborators who contributed with experimental work, calculations, and model developments.

# Author

**Guillermo Hough, Ph.D.,** is a research scientist with the Comisión de Investigaciones Científicas and works in the Instituto Superior Experimental de Tecnología Alimentaria in the small town of Nueve de Julio, Provincia de Buenos Aires, Argentina.

Dr. Hough conducts his research in the area of sensory science and has become an expert in the field of sensory shelf life, having published 20 refereed articles on this theme. His courses on sensory science and shelf life have been very popular in Argentina, Chile, Colombia, Italy, México, New York, Paris, and Peru. His group is in close contact with local and regional food companies, performing a wide range of sensory tests and panel training.

# chapter 1

# Introduction

## 1.1 Sensory shelf life definition

Different sources provide different definitions of *shelf life*. For example Wikipedia (http://en.wikipedia.org/wiki/Shelf_life, accessed September 5, 2008) says that "shelf life is the time that products can be stored, during which the defined quality of a specified proportion of the goods remains acceptable under expected (or specified) conditions of distribution, storage and display. Shelf life is different from expiration date; the former relates to food quality, the latter to food safety. A product that has passed its shelf life might still be safe, but quality is no longer guaranteed." Another definition was given by Fu and Labuza (1993): "The shelf life of a food is the time period for the product to become unacceptable from sensory, nutritional or safety perspectives." Ellis (1994) defined shelf life of a food product as "the time between the production and packaging of the product and the point at which it becomes unacceptable under defined environmental conditions." The IFST Guidelines (1993) defined shelf life as the time during which the food product will (a) remain safe; (b) retain desired sensory, chemical, physical and microbiological characteristics; and (c) comply with any label declaration of nutritional data, when stored under the recommended conditions. This last definition identifies the key factors that must be considered when assessing shelf life. The previous definitions are of shelf life as a whole, that is, covering to a lesser or larger degree microbiological, physicochemical, nutritional, and sensory aspects. The ASTM E2454 Standard (2005) defines sensory shelf life (SSL) as: "… the time period during which the product's sensory characteristics and performance are as intended by the manufacturer. The product is consumable or usable during this period, providing the end-user with the intended sensory characteristics, performance, and benefits."

The shelf-life limits of some of the factors mentioned in the definitions are defined at a laboratory level without the intervention of the consumer. For example, a central issue in declaring a food to be safe is that it must be free of pathogenic bacteria and this can be assessed by standard microbiological analysis. Another example is vitamin D–enriched milk, which must comply with a certain concentration measured in an analytical laboratory. In the shelf-life definitions the limit to sensory properties is referred to as acceptable/unacceptable or desired or intended. These limits can be highly ambiguous if not thoroughly researched. Manufacturers who have to comply with food regulations and, more importantly, comply with quality standards, need practical and reliable methods to estimate the shelf lives of their products.

## 1.2  Labeling regulations

Regulations on labeling of shelf-life information on food products vary from country to country. In Argentina (Código Alimentario Argentino 2005) it is compulsory, though what the label actually says is quite flexible. Manufacturers can choose from *use by*, *best before*, or *expires*. Some products are exempted from shelf-life labeling: wine, beverages with an alcoholic content above 10% (v/v), bread that is usually sold and consumed within 24 hours, vinegar, sugar, sweets, and salt.

In the United States (USDA 2008), except for infant formula and some baby food, product dating is not generally required by federal regulations. If a calendar date is shown, immediately adjacent to the date must be a phrase explaining the meaning of that date such as *sell by* or *use before*. There is no uniform or universally accepted system used for food dating in the United States. Open dating (use of a calendar date as opposed to a code) is found primarily on perishable foods such as meat, poultry, eggs, and dairy products. Coded dating might appear on shelf-stable products such as cans and boxes of food. Because the expiration date is not indicative of product quality if storage conditions have been less than optimal, the Food and Drug Administration (FDA) does not require expiration dates on most products. An exception to this is that expiration dates are required on drugs. The dates required on infant formula products are *use by* dates, not *expiration* dates. A consumer using the infant formula product before this date is assured that the product meets nutritional and quality standards (http://www.cfsan.fda.gov/~dms/qa-ind7c.html, accessed April 13, 2009).

In the United Kingdom (Food Standards Agency 2008) it is compulsory for food products to have "an appropriate durability indication." There is also a clear distinction between *use by* and *best before* dates. *Use by* means that any food or drink should not be consumed after the date shown on the label. Even if it looks and smells fine, using it after this date could put health at risk. *Use by* dates are usually placed on foods that go bad quickly, such as milk, soft cheese, ready-prepared salads, and smoked fish. *Best before* dates are usually used on foods that last longer, such as frozen, dried, or canned foods. It should be safe to eat food after the *best before* date, but the food will no longer be at its best. After this date, the food might begin to lose its sensory quality.

No doubt other countries, or states within countries, have varying regulations on shelf-life labeling. Gone are the days when food was bought at local stores like the butcher shop, produce store, or grocery store; or even sold door to door as with local milk products. In those days there was a general trust in the products' freshness. Today more and more consumers expect shelf-life dating to be on packaged food products.

## 1.3   Shelf life of foods is sensory shelf life

Some of the variables that must be considered when dealing with shelf life are as follows: the nature of the food, its composition, the ingredients, the processing it went through, the packaging used for its protection, storage conditions, distribution, and handling—both by retailers and the consumer. It is well known that these factors can have a negative influence on the quality attributes of a food product.

There is no doubt that the most important aspect that needs to be guaranteed during shelf life is sanitation. No food company can tolerate customer illness from pernicious microbial growth or from the presence of a toxic chemical substance developing during prolonged storage. That is, the first step is to make sure consumers will come to no harm through eating a food during its established shelf life.

For some foods the nutrition aspect is crucial. For example, in human milk replacement formulas, vitamins and other essential nutrients should not suffer destruction during storage. Loss of essential amino acids due to the Maillard reaction can also be an issue during storage of some dairy products such as evaporated or powdered milk.

Once the sanitary and nutritional hurdles have been overcome, the remaining barrier depends on the sensory properties of the product. It could be argued that chemical or physical changes have to be considered, and this is true for a full understanding of the deterioration process, but they directly affect sensory quality. For example, during storage of set-type yogurt there is a physical phenomenon, the contraction of the gel, and this liberates a milky-looking liquid that is not easily seen by the consumer. During the storage of a fruit juice chemical reactions take place, which lead to browning. Mechanisms can be sought that delay or suppress these chemical reactions, but what these mechanisms are really after is avoiding a sensory property that has a negative impact on consumer perception.

An observation of the aisles of the food and beverage sections of a local supermarket in Argentina showed that foods can be classified as having their shelf life determined by sensory, microbiological, or nutritional properties:

- *Sensory properties:* alcoholic drinks (beer, wine, cider, whiskey, etc.), carbonated soft drinks, fruit juices, energy drinks, water, concentrated beverages (liquid or powdered), sterilized milk, flavored sterilized milks, yogurt, butter, margarine, processed cheese, semi-hard cheese, hard cheese, grated cheese, powdered milk, artificial sweet-eners, coffee, cocoa, teas, jams, breakfast cereals, biscuits, snacks, dry pasta, flour, bread crumbs, cake mixes, flour mixes, corn starch, oats, rice, dried legumes, mayonnaise, ketchup, mustard, sweets, pow-dered custards, powdered jellies, olives, canned products (meats, fish, vegetables, fruits, etc.), chocolate, candy, dry soups, oil, frozen products, bread, cakes, cured meat products (raw ham, salami, etc.), spices, flavorings, fresh vegetables, and fresh fruits.

- *Microbiological properties:* pasteurized milk, soft cheeses, fresh dairy desserts (custard, mousse, etc.), eggs, cooked meat products (bologna sausage, Vienna sausages, etc.), fresh meats, and fresh pasta products (ravioli, gnocchi, etc.).
- *Nutritional properties:* baby infant formulas.

In other supermarkets, countries, or cultures the above list can change, but the number of food products whose shelf life is dependent on their sensory properties is far greater than those products whose shelf life depends on microbiological and/or nutritional properties. Even some foods that were placed in the microbiological category may have their shelf life determined by sensory properties, depending on the manufac-turing process and storage condition. This is the case of bologna sausage that takes on a dry appearance if cooked in an autoclave and packaged in a permeable film before the onset of microbiological failure. Thus, when talking about shelf life of foods, in the vast majority of cases we are talk-ing about *sensory* shelf life.

## 1.4   Importance of the consumer in defining food quality

A total quality approach for a manufactured food product will include (Ellis 1994):

- Product design (including hazard design analysis and risk assess-ment to ensure safety)
- Specification and testing of ingredients and packaging materials
- Manufacturing processes
- Transport, storage, and retail display
- Storage at home and consumption

Since the food must be safe and have an acceptable quality when con-sumed, the time for which this is maintained—the shelf life—is an essen-tial aspect of product design. How is the acceptable quality defined? Who is to answer this question? In the shelf-life definitions given above the limit to sensory properties was referred to as acceptable/unacceptable or desired or intended. More general definitions of quality (Muñoz et al. 1992) all mention terms like "ability to satisfy given needs," "fitness for use," or "conformance to requirements." It is not a coincidence that shelf-life definitions have items in common to quality definitions. Quality control and shelf-life tests are often similar; and many of the changes in aging of a product are also quality problems, such as deterioration of texture, browning, oxidation, syneresis, and off-flavor development (Lawless and Heymann 1998). Whether referring to shelf life or a more general quality approach, the question arises as to who decides whether a product is acceptable or whether it has the ability to satisfy given needs?

Attention should be paid to shelf-life issues throughout the food chain, always bearing in mind that consumers are the final link in that chain and in the end they will judge whether the product is acceptable and/or satisfies their needs. Above, attention was drawn to the fact that the majority of current food products have their shelf life defined by sensory properties, and these are the properties that can and will be evaluated by consumers.

The ASTM E2454 Standard (2005) on shelf life gives three possible criteria to define the shelf-life end-point: (1) change in the overall sensory profile, (2) a product's attribute(s) (including off-notes) that is (are) known or suspected to be key to the consumers' perception that the product has changed, and/or (3) consumers' considering the product no longer acceptable. Companies that are very stringent in their standards might choose the first criterion for which discrimination or descriptive tests are necessary, and these will be discussed in Chapter 2. For the second criterion consumer input is necessary to identify critical attributes, and once these have been identified, a trained panel can be used to monitor their changes. The third criterion is valid for those products whose manufacturers are willing to accept that the stored products are not identical to the fresh products. In this case consumers must identify the magnitude of change they are willing to tolerate. As pointed out by Cardello (1998), although an in-house panel may find a significant sensory difference between a production sample and a control, or between a sample stored for 1 month at 35°C and a control stored at 5°C, this does not necessarily mean that consumers will notice the sensory change; and if they do notice the change, they may disregard it as unimportant and happily consume the product as usual.

In Chapter 6 the cut-off point methodology will be presented where the consumers define limits to sensory defects generated during storage. Sensory shelf-life articles are frequently published in which the end-point has been arbitrarily defined by a researcher. Escalona and others (2007) studied changes of fresh-cut kohlrabi sticks during storage under modified atmosphere packaging. To determine sensory shelf life a five-member panel used a 9-point scale ranging from 1 = *inedible* to 9 = *excellent*, with 5 = *fair* considered the "limit of marketability." No information was given as to how this limit was decided on. Martínez and others (2006) researched the combined effects of modified atmosphere packaging and antioxidants on the storage stability of pork sausages. A six-member trained panel used a 5-point scale to measure off-odor. The scale went from 1 = *none* to 6 = *extreme* with 3 = *small*, the last score considered as a limit of acceptability in reference to an article by Martínez and others (2005), who in turn traced this limit of acceptability to Djenane and others (2001). In this last paper the 1–5 scale was used by a six-member, trained panel, but there was no mention of 3 = *small* corresponding to a limit of acceptability. Thus the sequel ends on a lost trail. Martínez and others (2006) stated that their treatment extended the shelf life of pork sausages from 8 to 16 days based on this limit of 3 on their 1–5 scale. What a consumer would think when eating these pork sausages was apparently never considered.

## 1.5   Books on shelf life of foods

This section covers the most relevant books on shelf life of foods published in the English language.

### 1.5.1   Labuza (1982)

The pioneering book on shelf life of food was written by Labuza (1982). The first chapter is an interesting introduction as to how open dating of foods came about. The positions of the different actors involved in the issue are presented: consumers, industry, and government. Consumers across the United States strongly supported the concept of putting open dates on food products as a right to know the age of the food they buy. At the time the food industry was divided over the question of open dating. Retailers were in favor, manufacturers of semi-perishable and non-perishable food items were against any form of open dates, and manufacturers of perishable goods were in favor of some form of open dating. State government officials had divided opinions as to mandatory laws on open dating.

Chapter 2 of Labuza's (1982) book outlines basic food degradation and preservation issues. The following is a list of the degradation modes briefly treated:

- *Preharvest biological decay:* undesirable changes occurring prior to harvest or slaughter shorten future shelf life; processing does not make low-quality foods better.
- *Senescence:* ripening of fruit after harvest or aging of meat after slaughtering is a desirable change; however, eventually enzymatic

biochemical processes lead to degradation, including loss of color, flavor, nutrients, and texture.

- *Microbiological decay:* microorganisms constitute a major mechanism by which many foods, especially minimally processed ones, lose their quality. Most food processes are designed to guard against the presence of pathogens. Inappropriate handling of the food by the consumer can lead to post-contamination, even within the *use by* date. It is thus important for the consumer to receive adequate education and instructions on proper food handling.
- *Chemical deterioration:* enzymatic, lipid oxidation, non-enzymatic browning, light-induced reactions, and oxidation of vitamins come under this heading. Lipid oxidation is a particularly common mode of deterioration. When consumers taste a product that has undergone either lipid oxidation or lipid hydrolysis, they may say the product is rancid. Chemical analysis or a panel trained in detecting lipolysis flavor as different to oxidized flavor can inform the food producer what the consumer was really detecting. Very little fat has to oxidize for the consumer to detect rancidity. Chemical deterioration reactions produce changes in sensory properties that can be perceived by the consumer.
- *Physical degradation:* bruising/crushing, wilting, moisture gain/loss, temperature-induced texture changes, and staling of flour products such as bread. As with chemical deterioration, these physical changes are perceived by the consumer through the senses. These changes do not affect the safety or nutritional properties of the food; a bruised fruit or hardened toffee is in no way dangerous to eat but will be rejected by the consumer due to its poor sensory quality.

The basic food-processing principles are also treated in Chapter 2 of Labuza's (1982) book:

- Heat is used to (1) destroy pathogens, (2) destroy spoilage microbes, (3) denature enzymes, and (4) soften tissues to make the food more digestible.
- Cold preservation is obtained by refrigerating and freezing. Although the general rule is the lower the temperature, the slower the rate of deteriorative reactions, this rule has its exceptions. For example, freezing some foods just below the freezing point actually increases degradation rates of some reactions. This is because not all water is frozen out and the chemicals are concentrated in the remaining water.
- Water content can be controlled by concentration, drying, use of humectants such as sugars, extrusion (combination of heat and drying), and deep-fat frying, which is basically a drying process.

- Chemical preservation can be obtained by fermentation, one of the most ancient food processes. In this method, a desirable organism is allowed to partially convert some of the carbohydrates of the food into acid, alcohol, and flavor compounds. The acid and/or alcohol prevent the growth of pathogenic microorganisms. In some cases further processing is done, such as pasteurization of beer or refrigeration of yogurt. Other modes of chemical preservation are through the use of additives such as acids, smoking agents, antioxidants such as BHA, chelating agents such as EDTA, or metabolic inhibitors like sodium benzoate.
- Controlled atmosphere storage is a mode of chemical preservation used mainly to extend the shelf life of fruits and vegetables. The rate of respiration after harvest is reduced by reducing the availability of oxygen or by increasing $CO_2$ levels. The methods can be classified as (1) reduction and control of $O_2$ by use of nitrogen, (2) hypobaric storage by pulling a vacuum, and (3) use of both increased nitrogen and increased $CO_2$ in the head space.

Chapter 3 of Labuza's (1982) book is titled "Scientific Evaluation of Shelf Life." An introductory paragraph informs about the work done by the U.S. Army in the late 1940s through 1953 to gain information for the military food supply. Basically for each food item, each mode of deterioration was studied at several temperatures for up to 3 years. The results of this study included many reports and tables. For example, a table would indicate that sliced bacon packaged in a can could be kept for 48, 18, and 12 months at 40°F, 70°F, and 90°F, respectively. However, as Labuza stated, much of this information is inapplicable today because the foods are processed differently, different packaging systems are used, and the distribution system has changed. It could be added that the consumer has also changed, generally becoming used to better quality foods than those consumed 60 years ago.

The rest of this chapter sets down the basic food deterioration kinetic equations. Zero- and first-order reaction rate equations are presented and discussed. The well-known Arrhenius relationship of reaction rate versus temperature is presented, together with the $Q_{10}$ concept. For the case in which storage temperature cannot be considered, constant prediction equations for variable temperature conditions are developed. Moisture effects on shelf-life predictions are also covered, with a thorough discussion on how to obtain accelerated conditions for moisture uptake or moisture loss during storage. Since then, no other book has treated the subject of moisture effects so thoroughly.

Following these initial basic chapters (Labuza 1982), Chapters 4 through 21 deal with specific food categories. For example, Chapter 4 covers cereal grains and flour; Chapter 11, dairy products; and Chapter 21, fresh fruits and vegetables. The issues covered in each chapter are mainly the principal modes of deterioration and shelf-life data available at the time.

As an example, the contents of Chapter 11 on shelf life of dairy products will be discussed. As an introduction processing and modes of preservation of different dairy products are summarized; packaging and distribution are also described. Following are the modes of deterioration. On page 199 an interesting table presents flavor and odor defects and their probable causes. This table is general but gives initial clues as a basis for further investigation. The main part of the chapter relates to shelf-life data based on an extensive review of data available at the time. Some of the data are confusing. For example, for pasteurized homogenized milk, activation energies for flavor deterioration range from 17.7 kcal/mol to 31 kcal/mol. Shelf life for this product at 40°F read from two different graphs was between approximately 6 and 25 days. Data based on the U.S. Army studies surprisingly have $Q_{10}$s all around two, with no given mode of deterioration. Interesting data is presented for nonfat dry milk with different moisture content where large differences in flavor activation energy were probably due to different deterioration mechanisms at different moisture contents. Overall, if a present-day company is to conduct a sensory shelf-life test on a dairy product, it would be wise to consult this chapter as an initial orientation. Lack of precise details of deterioration, conflicting data, and outdated production and packaging methods prevent the data from being used without further research.

## 1.5.2  Gacula (1984)

The next in chronological order in the published books on shelf life of foods is a chapter written by Gacula (1984). The two main innovative concepts introduced by Gacula (1984) are (1) food shelf-life studies are censored times and (2) the life of an item is assumed to be governed by a probability distribution.

*Censoring* refers to uncertainty in the exact failure time. For example, when an experiment is to be run for only a designated length of time, not all items in the experiment will have failed during this time. What is known is that the item lasted until the experiment ended. In Chapter 4 I will introduce survival analysis concepts to be used in sensory shelf-life estimations. My focus will be on the probability of consumers rejecting a product and not on the product's failing per se, which is the focus given by Gacula (1984).

In this chapter failure was decided upon by arbitrarily assigning a cut-off point on a sensory scale. In one example an off-flavor scale was used with 1, *no off-flavor* and 7, *very strong off-flavor*. An average score from

10 trained panelists of 2.5 or more was considered high enough for the product to have failed. No consumer input was used to define this cut-off point, as shall be presented in Chapter 6.

The following parametric distributions were presented by Gacula (1984): normal, log-normal, and Weibull. Based on the Weibull distribution he presented a detailed example of how shelf life can be estimated using hazard plots based on the notion of exact failure times and right-censored times. In Chapter 4 I show that in sensory shelf-life testing, interval-censoring is the most frequent type of censoring encountered. This type of censoring was not considered by Gacula (1984).

Finally Gacula (1984) addressed the regression of sensory scores versus storage times as a means of estimating shelf life. With a practical example he presents the regression equations to estimate a straight-line regression together with its confidence intervals to thus be able to obtain confidence intervals for the shelf-life estimation. This methodology will be presented in detail in Chapter 6.

## 1.5.3   IFST guidelines (1993)

The food shelf-life definition given by the IFST (1993) in its introductory chapter has already been cited as one that identifies the key factors that must be considered when assessing shelf life. Another interesting concept presented in the introduction is related to the importance of consumers in defining shelf life: "… as the food is being produced to be consumed it is right that first consideration should be to the importance of shelf life to us all as consumers." In future chapters of this book this importance of the consumer in defining sensory shelf life will be stressed. The rest of the introduction considers the different actors who play a role in shelf life during food manufacture, from growers to retailers.

Chapter 2 of the IFST guidelines (1993) refers to the general factors influencing the shelf life of foods. Various factors such as raw materials, formulation, processing, hygiene, packaging, and storage are briefly described. One factor mentioned is consumer handling: what happens between point of purchase and consumption remains very much an uncertain factor as little published information covered this issue. Since then, little has changed, and there is still a need for published information on how consumers store and handle food between purchase and consumption. Chapter 3 deals with mechanisms of deterioration in a similar manner as described in Chapter 2 of Labuza (1982).

Chapter 4 covers different phases in shelf-life determinations: initial, preliminary, confirmatory, and routine shelf-life determinations are briefly defined. The setting up of a shelf-life study is presented with considerations regarding samples and sampling during storage. Shelf-life tests are classified as sensory evaluation, microbiological examination, chemical analysis, and physical examination. They state that sensory evaluation tends to be the most common method of test because it is easier to carry out in-house than, for example, chemical analysis. This statement is arguable because the know-how and facilities to carry out good quality sensory evaluation are lacking in many food companies that have sophisticated physicochemical laboratories. The IFST guidelines (1993) do not involve the consumer in shelf-life studies. All the recommendations concern trained panels performing difference testing and profile analysis.

Chapter 5 refers to accelerated shelf-life studies that constitute an "indirect approach." They stress the limitations of these studies: (a) cost of conducting accelerated studies to ensure statistical confidence in the results; (b) as temperature rises, a change in physical state may occur; (c) moisture loss can occur when accelerated temperature conditions are used, thus leading to false results; (d) upon freezing, reactants are concentrated in the unfrozen liquid, resulting in unpredictable, high reaction rates; and (e) different microorganisms grow at different temperatures. Predictive food microbiology is briefly discussed.

Chapter 6 deals with the necessity of conducting routine monitoring of product shelf life, especially as a means of checking that the shelf life set during development, possibly using an accelerated storage test, remains correct for the normal storage condition. Customer complaints and the use of time–temperature indicators are briefly discussed.

Overall the IFST guidelines (1993) comprise a compound manual of main issues concerning shelf-life studies of food.

## 1.5.4   Man and Jones (1994)

The first five chapters (Part 1) and the next ten chapters (Part 2) come under the headings of "The Principles" and "The Practice," respectively; however, there is some overlap. There is no specific chapter or section on sensory shelf life.

Chapter 1 (Singh 1994) presents major modes of deterioration, kinetic equations, and the types of sensors used to monitor shelf life of foods. Among the kinetic equations a non-linear approach to calculating activation energy is suggested, although the true value of this approach, which consists of more precise estimation of the activation energy, is not discussed. Chapter 7 of this book will show that this is the case in sensory shelf-life estimations.

Chapter 2 (Ellis 1994) introduces an arguable concept:

> Expert or consumer panels producing statistically analyzed results may give the best measure of storage changes and acceptability but can be very expensive for regular use. An individual shelf life

assessor with proper training and experience of
normal storage changes may be employed to carry
out most assessments and other resources such
as analytical services or expert taste panels used
where necessary.

This proposal of replacing sensory panels by an individual "shelf life
assessor" has the obvious drawback of relying on the judgment of a single
individual. Thurstonian theory (Lawless and Heymann 1998) indicates
that the strength of a perceived sensory stimulus is normally distributed,
basically meaning that an identical stimulus presented to the same individual is sometimes perceived weak, other times strong, and other times
in between. Thus, this proposal by Ellis (1994) of relying on the judgment of a single individual would lead to samples rejected incorrectly
(the single assessor perceived an acceptable off-flavor as overly strong), or
samples accepted when they shouldn't be (the single assessor perceived
an objectionable off-flavor as weak). Having multiple responses from a
panel of trained assessors neutralizes the effect of extreme judgments
from a single assessor. Having a single assessor decide whether a sample
is acceptable or not, thus having him/her replace a consumer panel, is not
recommended at all. A trained assessor's function is to measure the intensity of analytical sensory attributes, not to measure acceptability. Trained
assessors should not even form part of a consumer panel because they
have been taught to use an analytical frame of mind different from the
way a naïve consumer approaches a food judgment. In the same book
Eburne and Prentice (1994) stated:

> It is not always possible for formal taste panels to
> be used in shelf-life determinations and frequently
> it is left to the experience of a few individuals to
> determine whether a product is acceptable or not.
> This is only possible if the individuals carrying out
> the assessment are highly skilled with considerable
> experience of sensory analysis techniques.

In no way can a few skilled assessors predict whether a product will
be acceptable to consumers or not, and relying on this methodology
will inevitably lead to mistakes. Reilly and Man (1994) also presented
this same conceptual error when they propose that 8–10 trained assessors
should decide on the acceptability of stored potato crisps.

The central argument used by Ellis (1994) to disregard good sensory
practice is that it is expensive. This may be so, but if the result cannot
be achieved properly by other means, the discussion is whether to use
sensory analysis or not. If a vitamin assay is deemed too expensive, the

chemist does not propose replacing the correct assay by a cheaper, unreliable method. The decision is whether to perform the assay or not.

In Part 2, "The Practice" (Man and Jones 1994), different authors cover different food products, discussing product characteristics, factors affecting shelf life, its determination, and current developments. In their sections on shelf-life determination all authors recognize the importance of sensory evaluation, but only some go into details. Howarth (1994) introduces the concept of a cut-off point for breakfast cereals, whereby samples with different moisture content would be evaluated by trained assessors and the moisture level at which the product is not acceptable is determined. In Chapter 6 of this book the cut-off point methodology based on consumer and trained panel measurements will be presented. Howarth (1994) presents three case studies of cereal shelf-life determination with sensory evaluation considerations. In one of these cases it is stated that a month of accelerated storage is equivalent to a month's "ambient life." What accelerated or ambient conditions were and the basis of the time equivalence, are not indicated.

Goddard (1994) presents three approaches to determine sensory shelf life that imply answering one of the following questions:

1. *How long can a product be stored without noticeable changes in it sensory attributes?* This question is answered by discrimination tests and will be discussed in Chapter 2 of this book.
2. *How do sensory attributes change on storage?* This question is answered by descriptive analysis performed by a trained panel. The objective is to determine when a critical sensory attribute reaches a predetermined, unacceptable level. How this unacceptable level is chosen is not discussed in the chapter. The cut-off point methodology, which will be presented in Chapter 6 of this book, describes methodology to determine the unacceptable level based on consumer input.
3. *How long can the product be stored before changes in sensory properties render it unacceptable?* This implies working with consumer panels as will be discussed in this book in Chapters 4 and 5, where survival analysis methodology is presented.

Symons (1994) defined *practical storage life* as the time during which the product under test will still be appreciated by the ultimate consumer as meeting the level of quality expected for the product. The method proposed to measure practical shelf life was to have a trained panel determine the time to "just noticeable difference" and then multiply this by an arbitrary figure, "generally between 2 and 5." Using this wide range of arbitrary figures would lead to shelf-life values of poor reliability.

## 1.5.5   Taub and Singh (1998)

This book contains some interesting chapters. The importance these authors gave to sensory and consumer perception of food stability lies in the fact that the initial and final chapters of the book are devoted to these issues.

The first paragraph of Chapter 1 (Cardello 1998) indicates that consumers, through their purchase and/or nonpurchase of food products, must be considered the ultimate arbiters of food quality. Cardello made a distinction between intrinsic versus extrinsic factors affecting food quality. Among the first are ingredient, processing, and storage variables, which control the sensory characteristics of the product and are the most important variables determining both the acceptability and perceived quality of the item to the user. Although intrinsic factors are significant determinants of food quality, many factors that are extrinsic to the product also play a role. These factors include variables such as attitudes, expectations, and social or cultural influences. Different extrinsic factors are analyzed:

- Environmental factors: the physical environment (for example, a hot or cold day) and the body's internal environment as affected by heat, cold, or physical exercise. The effect of these factors on sensory shelf life has not been investigated. For example, would hunger and/or thirst influence the perception of a sensory defect originated in prolonged storage?
- Learning and conditioning: very few sensory experiences are innately preferred or innately rejected. Most sensory-hedonic associations are learned. Food aversion studies have shown that a single pairing of food with illness was sufficient to establish a strong and prolonged aversion to the food. The pairing of a product with a poor sensory experience due to consumption of the product after prolonged storage might affect future acceptability of this product or a product under a particular brand. This issue has not been addressed in published research.
- Expectation and contextual effects: a food stimulus occurs within a contextual background of other stimuli, past experiences, and perceptions that establish an expectation or baseline against which the stimulus is judged. The assimilation model has generally been proved to apply in expectation studies. This model predicts that (1) product information, packaging, or advertising appeal that increases the consumers' expectations for the product will increase its perceived acceptability and perceived quality, and (2) any negative associations with the product that lower expectations will result in decreased acceptance, regardless of "true" product quality. How these factors affect consumers' perception of sensory shelf life has received little

attention in published research. For example, is the shelf life given by a leading brand more credible to the consumer than the shelf life of less-known brands? Would this lead to the consumer's accepting higher levels of a sensory defect originated by prolonged storage in a leading brand than in a secondary brand? What happens when a consumer observes that the *best before* date is far away? Does this lead to a positive expectation and thus a greater tolerance to a sensory defect? Inversely, if the *best before* date is close by, does this lead to a negative expectation and consequently a higher rejection probability regardless of the objective sensory properties? Answering these questions would lead to a better understanding of how consumers perceive sensory shelf life.

- Social and cultural influences: religious or cultural taboos, such as avoidance of pork by Jews and Muslims and aversion to cat and dog meat in many countries. Household income also plays a role in food acceptability. Data presented in Chapter 5 of this book will show how children from a low-income population had a lower rejection probability for stored yogurt than children from a medium-income population. It has been shown that parents, peers, leaders, and heroes have an influence on food acceptability. Also the total amount eaten in a meal increases when the meal is eaten with others as opposed to when it is eaten alone. Do these factors affect sensory shelf life in the sense that if a consumer sees that all those around him are happily eating a product that he has perceived as having a sensory defect, would he disregard the perception and eat the product anyway? This is an interesting point of research.
- Psychological and emotional factors: personality factors that, for example, may predispose a person to be obese, or the study of sensation-seeking individuals and how this may affect their food choices. Are such individuals more likely to consume a food close to its *best before* date than individuals more conservative in their food choices? This is another point of interesting research.

Cardello (1998) ended his chapter by making a point on the equivalence between food quality and overall liking: Consumers' perception of food quality is intimately linked with what they like. Chapter 4 of this book will show how consumers' acceptance/rejection of food products with different storage times can be used to predict sensory shelf life.

Chapter 11 of Taub and Singh's (1998) book (Ross 1998) presents basic kinetic equations related to shelf-life calculations not unlike presentations available in other books previously mentioned (Labuza 1982, Mizrahi 2000). In a section entitled "Statistical Determination of Shelf Life," Ross presented the classical two-step approach to accelerated testing calculations. This two-step approach implies calculating reaction rate constants at different temperatures (first step), and then using these reaction rates to estimate activation energy (second step). Ross did not consider the non-linear approach, which had already been presented by Singh (1994) as noted above when discussing Man and Jones' (1994) book. Chapter 7 of this book will show that for sensory shelf-life testing, a one-step approach using non-linear regression is more efficient from a statistical point of view.

Other chapters in Taub and Singh's (1998) book deal with physico-chemical aspects of food preservation. Chapter 13 (Wright and Taub 1998) addresses the issue of quality management during storage and distribution, which is not covered by other food shelf-life books. Yang (1998) introduced a conceptual error when he states that "accelerated testing at 38°C is based on the principle that reactions are reduced half by each drop of 10°C in temperature." This rule, based on the validity of a single activation energy valid for all food products, is false (Nelson 1990). The last chapter of the book (Bruhn 1998) broadly covers the subject of consumer attitudes and perceptions, but not from a shelf-life perspective.

## 1.5.6   Kilcast and Subramaniam (2000a)

This book contains a number of interesting chapters. The introductory chapter was written by these same editors (Kilcast and Subramaniam 2000b). This chapter defines shelf life in the terms of the IFST (1993) definition given above and points out that terms such as *desired characteristics* placed in a definition are highly ambiguous. The rest of this chapter deals with factors influencing shelf life, similar to Labuza's (1982) corresponding chapter. Different shelf-life measuring techniques are summarized: sensory, physical, chemical, and microbiological. The design of shelf-life experiments is original in the presentation of basic and reverse storage designs, which will be discussed in Chapter 3 of this book. Tables are presented in the Appendix to this chapter; deterioration mechanisms and limiting changes are listed for different food products. There are tables for fruit and vegetable products, meat and meat products, cereals and other dry products, beverages, and dairy products. It is interesting to note that of the 39 food products listed, 6 have their shelf life defined by safety considerations due to microbial growth, and for all the rest shelf life is limited by sensory changes.

Chapter 2 introduces the concepts of water activity and the glass transition temperature for predicting and controlling the stability of food systems focusing on microbial growth. Chapter 3 is devoted to the mathematical modeling of microbial growth.

Chapter 4 (Kilcast 2000) presents an overview of sensory evaluation methodology with some particular references to sensory shelf life.

He presented three possible categories of criteria that can be used for interpreting shelf-life data:

- *First detectable change or just-noticeable difference:* this change is measured using difference tests, assuming that a suitable reference sample is available. These tests usually report small differences that have little relevance to sensory quality as perceived by consumers.
- *Quantitative measures of relevant sensory attributes:* if these are used, a fixed level of change has to be adopted as a cut-off point to decide the end of the product's shelf life. This cut-off point methodology will be addressed in Chapter 6 of this book. An interesting concept presented by Kilcast (2000) is that the growth of a non-characteristic attribute (for example, rancidity in butter) is often more easily detected than decrease of a characteristic attribute (for example, diacetyl flavor in butter). This sounds reasonable, although I am not aware of published research that supports this idea.
- *Consumer acceptability:* in the present book this issue will be addressed in the use of survival analysis techniques (Chapter 4) and the cut-off point methodology (Chapter 6).

In Chapter 5 Mizrahi (2000) addressed the issue of accelerated shelf-life tests (ASLTs). He stated that ASLTs are applicable to chemical, physical, biochemical, or microbial processes. Although these processes are responsible for sensory changes in the product, it is often of interest to analyze how a sensory attribute like rancid odor changes under accelerated conditions; sensory changes were not addressed by Mizrahi (2000). The examples presented in the chapter are based on chemical deterioration. The *initial rate* approach is presented in the first place. This approach is the simplest technique for ASLT. The idea is to measure deterioration in actual storage conditions for a short time and then extrapolate the kinetics to longer storage times. For example, a properly bottled vegetable oil can have a sensory shelf life of 2 years. If a very sensitive method was available to measure the oxidation level after only 1 month of storage and the reaction rate was known (for example, zero-order reaction rate), then the product's shelf life could be estimated by extrapolating the 1-month deterioration till the accepted cut-off point. From a sensory point of view this initial rate approach would not be applicable because sensory assessors, no matter how well trained, would be unable to reliably detect those initial deteriorations.

The simplest and most commonly used method for ASLT is based on employing only a single factor to accelerate the deterioration process. A contradiction was expressed by Mizrahi (2000) in the sense that in ASLT the validity of the kinetic model is crucial to obtaining accurate shelf-life prediction; yet the validity of the model cannot be fully verified by the

ASLT procedure because the levels used for the accelerated factor do not include actual storage conditions. For example, if an ASLT is conducted with vegetable oil in the 35°C–45°C range in order to predict shelf life at 20°C, the validity of the Arrhenius model might be checked between 35°C and 45°C, but it is doubtful whether the same model and/or parameters are valid for the whole range between 20°C and 45°C. The knowledge of sound physical and/or chemical theory related to the deterioration can help in this case.

Another interesting issue discussed by Mizrahi (2000) was the accuracy of the extrapolated data. Figures illustrating the error in extrapolated data in relation to errors in calculated linear slopes are presented.

Multiple accelerating factors are presented as an approach that has the advantage that the total acceleration effect of using two or more factors is a multiplication of their effect, while the error is only the summation. An example of this approach was studied by the author (Mizrahi 2000) for the process of non-enzymatic browning of cabbage where the effects of moisture and temperature were combined. The disadvantage of using multiple acceleration factors is that the experimental design is cumbersome, especially when the factors interact with one another. A *no model* approach is also presented; a difficulty in this approach is that the acceleration ratio is very dependent on how small a fraction of the total acceptable extent of deterioration may be significantly determined. For sensory changes small fractions of the total acceptable deterioration are difficult to measure.

### 1.5.7   Eskin and Robinson (2001)

These authors have edited a book that is divided into Physical Factors, Chemical Factors, and Biochemical Factors. There is little mention of the estimation of sensory shelf life.

### 1.5.8   Labuza and Szybist (2001)

In the opening chapter, the title, "Open Dating of Foods," was defined as the practice of labeling a packaged food with a date that indicates when the product was packed, meant to be sold by, or meant to be used by. The authors defined *shelf life* as the end of consumer quality determined by the percentage of consumers who are displeased by the product. This definition is basically the one adopted in the present book.

Chapter 2 focuses on the benefits and potential disadvantages of open dating. Although benefits outweigh disadvantages, it is worthwhile to point out the disadvantages also. One is possible lack of temperature control; if a food is temperature abused, an open date is meaningless and, in fact, provides a false sense of security. This is a serious concern when food safety is an issue, but not too worrying when dealing with sensory shelf life. Another potential disadvantage could be the financial costs. Shelf-life tests, printing equipment, and enforcement costs are mentioned. Another cost would be sorting habits at the retail level whereby consumers tend to choose the youngest dates and thus increase food waste.

Chapter 3 gives examples of temperature abuse episodes between 1980 and 1998. Following this discussion a case is made for the use of time–temperature integrators. These are small, physical devices that are placed on food packages to measure the temperature history of a product and thus indicate changes at the end of shelf life. Different alternatives to these devices are covered.

Chapter 4 presents an interesting table on approximate shelf lives of food product categories. The table is divided in perishable, semi-perishable, and long shelf-life foods. Following is an example of the information for fried snack foods:

- Food product: fried snack foods
- Principal modes of deterioration: rancidity, loss of crispness
- Critical environmental factors: oxygen, light, temperature, moisture
- Average shelf life: 4–6 weeks
- Date most suitable of product: *sell by* or *best if used by*
- Additional information: home storage information such as *store in a cool, dry place*

This information is of value when designing a shelf-life study because it can help in choosing proper storage conditions and by providing a guideline as to the maximum storage time. Information contained in Chapter 5 can also be used when designing a shelf-life study. It accounts for current labeling practices followed by different food categories and companies; brands and manufacturers are named. One interesting example is the informative pamphlet offered by Dannon for its yogurts (these days the pamphlet would be replaced by a Web page), which states: "Although the yogurt will remain fresh for at least a week beyond this date when properly refrigerated, its flavor changes and the yogurt becomes more tart with prolonged refrigeration." As Labuza and Szybist (2001) pointed out, this sort of information is confusing and inconvenient to the consumers and the employees at the retail level.

Chapter 9 describes in detail a survey conducted by the authors on home practices regarding open dates and storage of perishable refrigerated foods. Although the survey was centered on food safety and not sensory issues, it does show how important it is to have knowledge on consumer perceptions and habits regarding food storage.

More than half of the contents of the book are appendixes dealing with legislation on open dating in the United States, the European Union, and Canada. These appendixes could be of value to food regulation agents in different parts of the world seeking to implement or improve food storage legislation.

## *References*

ASTM E2454 Standard. 2005. *Standard guide for sensory evaluation methods to determine the sensory shelf life of consumer products.* West Conshohocken, PA: American Society for Testing of Materials.

Bruhn, C. 1998. Consumer attitudes and perceptions. In *Food storage stability*, ed. I.A. Taub and R.P. Singh, Chapter 19. Boca Raton: CRC Press.

Cardello, A.V. 1998. Perception of food quality. In *Food storage stability*, ed. I.A. Taub and R.P. Singh, Chapter 1. Boca Raton: CRC Press.

Código Alimentario Argentino. 2005. Normas para la Rotulación y Publicidad de los Alimentos, Chapter V, article 6.6. Buenos Aires, Argentina: Administración Nacional de Medicamentos, Alimentos y Tecnología Médica.

Djenane, D., A. Sánchez-Escalante, J.A. Beltrán, and P. Roncalés. 2001. Extension of the retail display life of fresh beef packaged in modified atmosphere by varying lighting conditions. *Journal of Food Science* 66: 181–186.

Eburne, R.C., and G. Prentice. 1994. Modified-atmosphere-packed ready-to-cook and ready-to-eat meat products. In *Shelf life evaluation of foods*, ed. C.M.D. Man and A.A. Jones, Chapter 8. Glasgow: Blackie Academic and Professional.

Ellis, M.J. 1994. The methodology of shelf life estimation. In *Shelf life evaluation of foods*, ed. C.M.D. Man and A.A. Jones, Chapter 2. Glasgow: Blackie Academic and Professional.

Escalona, V.H., E. Aguayo, and F. Artés. 2007. Quality changes of fresh-cut Kohlrabi sticks under modified atmosphere packaging. *Journal of Food Science* 72: S303–307.

Eskin, N.A., and D.S. Robinson (ed). 2001. *Food shelf life stability*. Boca Raton: CRC Press.

Food Standards Agency. 2008. Food labeling. http://www.food.gov.uk/foodlabelling/ (Accessed September 5, 2008).

Fu, B., and T. Labuza. 1993. Shelf life prediction: Theory and application. *Food Control* 4: 125–133.

Gacula, M. Jr. 1984. *Statistical methods in food and consumer research*, Chapter 8. Orlando: Academic Press.

Goddard, M.R. 1994. The storage of thermally processed foods in containers other than cans. In *Shelf life evaluation of foods*, ed. C.M.D. Man and A.A. Jones, Chapter 13. Glasgow: Blackie Academic and Professional.

Howarth, J.A.K. 1994. Ready-to-eat breakfast cereals. In *Shelf life evaluation of foods*, ed. C.M.D. Man and A.A. Jones, Chapter 12. Glasgow: Blackie Academic and Professional.

IFST. 1993. *Shelf life of foods: Guidelines for its determination and prediction.* London: Institute of Food Science & Technology.

Kilcast, D. 2000. Sensory evaluation methods for shelf-life assessment. In *The stability and shelf-life of food*, ed. D. Kilcast and P. Subramaniam, Chapter 4. Cambridge: Woodhead Publishing Limited.

Kilcast, D., and P. Subramaniam (ed). 2000a. *The stability and shelf-life of food.* Cambridge: Woodhead Publishing Limited.

Kilcast, D., and P. Subramaniam. 2000b. Introduction. In *The stability and shelf-life of food*, ed. D. Kilcast and P. Subramaniam, Chapter 1. Cambridge: Woodhead Publishing Limited.

Labuza, T.P. 1982. *Shelf-life dating of foods*. Westport, CT: Food & Nutrition Press.

Labuza, T.P., and L.M. Szybist. 2001. *Open dating of foods*. Trumbull, CT: Food & Nutrition Press.

Lawless, H.T., and H. Heymann. 1998. *Sensory evaluation of food, principles and practices*, Chapter 5 and page 557. New York: Chapman & Hall.

Man, C.M.D., and A.A. Jones (ed). 1994. *Shelf life evaluation of foods*. Glasgow, UK: Blackie Academic and Professional.

Martínez, L., I. Cilla, J.A. Beltrán, and P. Roncalés. 2005. Effect of different concentrations of carbon dioxide and low concentration of carbon monoxide on the shelf-life of fresh pork sausages packaged in modified atmosphere. *Meat Science* 71: 563–570

Martínez, L., I. Cilla, J.A. Beltrán, and P. Roncalés. 2006. Combined effect of modified atmosphere packaging and addition of rosemary (*Rosmarinus officinalis*), ascorbic acid, red beet root (*Beta vulgaris*) and sodium lactate and their mixtures on the stability of fresh pork sausages. *J. Agric. Food Chem.* 54: 4674–4680

Mizrahi, S. 2000. Accelerated shelf-life tests. In *The stability and shelf-life of food*, ed. D. Kilcast and P. Subramaniam, Chapter 4. Cambridge: Woodhead Publishing Limited.

Muñoz A., G.V. Civille, and B.T. Carr. 1992. *Sensory evaluation in quality control*, Chapter 1. New York: Van Nostrand Reinhold.

Nelson, W. 1990. *Accelerated testing. Statistical models, test plans and data analyses.* New York: John Wiley & Sons.

Reilly A., and C.M.D Man. 1994. Potato crisps and savoury snacks. In *Shelf life evaluation of foods*, ed. C.M.D. Man and A.A. Jones, Chapter 10. Glasgow: Blackie Academic and Professional.

Ross, E.W. 1998. Mathematic modeling of quality loss. In *Food storage stability*, ed. I.A. Taub and R.P. Singh, Chapter 11. Boca Raton: CRC Press.

Singh, R.P. 1994. Scientific principles of shelf life evaluation. In *Shelf life evaluation of foods*, ed. C.M.D. Man and A.A. Jones, Chapter 1. Glasgow, UK: Blackie Academic and Professional.

Taub, I.A., and R.P. Singh (ed). 1998. *Food storage stability*. Boca Raton: CRC Press.

Symons, H. 1994. Frozen foods. In *Shelf life evaluation of foods*, ed. C.M.D. Man and A.A. Jones, Chapter 15. Glasgow: Blackie Academic and Professional.

USDA. 2008. Food product dating. United States Department of Agriculture: http://www.fsis.usda.gov/Factsheets/Food_Product_Dating/index.asp (accessed September 5, 2008).

Wright, B.B., and I.A. Taub. 1998. Quality management during storage and distribution. In *Food storage stability*, ed. I.A. Taub and R.P. Singh, Chapter 13. Boca Raton: CRC Press.

Yang, T.C.S. 1998. Ambient storage. In *Food storage stability*, ed. I.A. Taub and R.P. Singh, Chapter 17. Boca Raton: CRC Press.

# chapter 2

# Principles of sensory evaluation

## 2.1 Introduction

The objective of the present chapter is to provide an introduction to sensory analysis principles and methodology, together with comments as to how this methodology relates to sensory shelf life (SSL). Those interested in pursuing the study of sensory analysis are encouraged to read specific textbooks of which a number are recommended (Lawless and Heymann 1998; Stone and Sidel 2004; Meilgaard et al. 2007). Readers with a background in sensory analysis who wish to move on to the specific issues of this book, might want to skip this chapter; however, before they do so they might want to look at Section 2.7.1.2, which addresses discrimination tests to determine shelf life and which will not be covered elsewhere in this book.

## 2.2 Definition of sensory evaluation

Stone and Sidel (2004) cite the definition given by the Sensory Evaluation Division of the Institute of Food Technologists:

> Sensory evaluation is a scientific discipline used to evoke, measure, *analyze and interpret* reactions to those characteristics of foods and *other materials* as they are perceived by the senses of *sight*, smell, taste, *touch* and *hearing*.

Following is a discussion of the words that I have italicized in this definition.

### 2.2.1 Analyze and interpret

Analyzing and interpreting relate to statistics, and I would say that 50% or more of the science sustaining sensory evaluation is statistical science. Some people just do not like statistics or are cagey about believing in its methodology; coping with sensory evaluation will be a hard task for people thus inclined. As will be seen in future chapters of this book, statistics play a fundamental role in predicting SSL. One aspect that is not contained in the above definition and is related to statistics is *design*. The first part of the definition could read: "Sensory evaluation is a scientific discipline used to *design*, evoke, measure, analyze and interpret..." Statisticians are forever warning practitioners to design their experiments with anticipation to ensure maximum efficiency and analyzable results. Taking this advice is particularly necessary in SSL studies, as will be discussed in Chapter 3. For example, the simple exercise of calculating the number of samples to be stored at a desired temperature is fundamental to ensure sufficient material is available for trained panels, consumer panels, physicochemical and/or microbiological studies. Very often, SSL studies can extend for weeks or months. If careful and proper designs are not considered, the experiment can come to an end with no meaningful results; or halfway through the study the experimenter suddenly realizes that he is running out of control samples due to not having considered that at each time point a control and blind control are necessary for correct sensory evaluation.

## 2.2.2   Other materials

Sensory evaluation was originally developed for food products but was later adopted for diverse products and situations. Tobacco, textile, automotive, and home-use industries are some products that rely on sensory evaluation to develop and control quality. The focus of this book is on SSL of food products; however, some cigarette brands in Argentina have a *best before* date on them, and the prime consideration in deciding on this date is most surely sensory. Textile products don't have a "shelf life" because they do not deteriorate under normal storage, yet estimating how long a pair of jeans lasts under normal use is a problem not very different from estimating how long a pot of mayonnaise can be on a shelf before being rejected by a consumer. Also, even when dealing with food products, other materials such as the packaging are involved in SSL studies. For example, bottled water exposed to high temperatures or intense light develops flavors associated with the plastic bottle. Keeping track of these off-flavors over storage time requires a panel trained in their recognition.

## 2.2.3   Sight, touch, and hearing

Taste and smell are what we first think of when referring to sensory analysis. Moskowitz and Krieger (1995) in a study of six food categories concluded that on average the rank of importance for attribute liking was taste/flavor, texture, and appearance; however, this tendency was by no means uniform over individuals, and the dispersion over individuals depended on the product. In Chapter 6 an example of a study on the SSL of a fluid human milk replacement formula will be presented in which the critical descriptor was the dark color that developed during storage. Our sense of touch encompasses a number of different sensations such as hardness, crispness, creaminess, or smoothness. Crispness and/or crunchiness are critical descriptors in defining the SSL of a number of products such as biscuits or snacks and are evaluated both by touch (mechanical breakdown between our fingers or in the mouth) or hearing while chewing the product. The SSL of French-type bread, an everyday staple food in many countries, is defined by its change in texture measured by our sense of touch. Dry-cured sausages of the salami family are sold in grocery stores and butcher shops in countries such as Spain or Argentina. Usually the different varieties are hung above the counter, and clients test their shelf lives by squeezing them between their thumbs and index fingers—hardly a hygienic practice, but the skin is not eaten. The meat might be judged as okay, undermatured (too soft), or overmatured (too hard). A combination of sight and touch can be found when consumers judge texture visually. For example, reconstitution of aged milk powder can produce badly dissolved granules, leading to a rejection of the product by the consumer. Another example is the visual evaluation of grating cheese fracturability when observing the whole cheese cut/cracked in half (Hough et al. 1994). Amerine et al. (1965) commented that degree of liking and flavor quality are synonymous in the minds of many consumers; these consumers do consider whether the cause of their liking or disliking the product was in fact appearance, texture or flavor.

## 2.3 Sensory analysis: Trained panels versus experts

In the previous section the definition of sensory evaluation was presented and some of its implications discussed. When we refer to sensory analysis we are referring to a methodology based on measuring sensory properties by panels of trained assessors or consumer panels. Experts have been used traditionally to assess some products such as wine or tea. Table 2.1 shows the major differences between sensory analysis as performed by a trained panel and expert methodology.

*Table 2.1* Differences between Sensory Analysis by a Trained Panel and by Experts

| Trained panel | Experts |
|---|---|
| Measure with their senses using standardized language | Say what they feel using subjective language |
| Assessors are screened and trained | Chosen by experience |
| A trained panel doesn't measure acceptability | Confuse quality control with hedonics |
| Performance of assessors is monitored | No control over their performance |

In consulting one of the many Web pages dedicated to wine characteristics, I came across *tired*, a term used by experts to somehow define a
wine having been subjected to prolonged shelf life (http://www.lovewine.
org/; accessed December 2008). The definition of the term was not very
helpful: "worn-out, past its prime, describing a wine that is fading." It is
possible that a wine expert can distinguish between a wine at its prime
and one that has been stored for too long, but what is not clear is how this
expert is going to transmit what he actually feels when using a subjec-
tive term such as *tired*. Compare this with *honey*, one of the descriptors
used by Presa-Owens and Noble (1997) when studying the effect of storage at elevated temperatures on aroma of Chardonnay wines. Should I
ask a panel of assessors if a stored wine has a higher honey note than a
control sample, they may answer that they are not sure what *honey* means
in wine. In this case I will prepare the honey reference as indicated by
Presa-Owens and Noble (1997): 1 ml of honey (Aunt Sue's Old Fashioned
Country Style) in 50 ml of Chardonnay wine. This reference will help in
training the panel to measure the honey note in wine with their senses
using a standardized language as indicated in Table 2.1.

Some years ago our laboratory was asked to screen and train a panel
for a dairy industry. Till then sensory evaluation was carried out by one
person who had been with the firm for more than 20 years and whom
everybody had trusted as the sensory "expert." This person had been chosen due to the experience he had acquired during many years of tasting
and was obviously included as one of the candidates to form part of the
trained panel. In the first screening test, candidates had to have more than
80% correct answers in basic taste identifications to continue; the "expert"
had less than 60%. We would have excused him from continuing, but the
company asked to keep him in the run because he was the "expert." He
continued with odor identification and triangle tests and presented a truly
low performance. Screening tests are generally designed to detect people
who have normal sensory acuity; this person had a sensory problem. This
was a real case that demonstrated the negative consequences of choosing
an expert due to his experience rather than his expertise.

In Chapter 6 (Section 6.3), details will be given on a study on the shelf
life of sunflower oil. In one part of this study, consumers tasted oils with
different degrees of oxidized flavor. Figure 2.1 schematically shows the
behavior of the consumers. The slope of acceptability versus oxidized flavor was negative for the majority of the consumers; that is, the fresh sample received the highest score, and the highly oxidized sample, the lowest
score. But 15% of the consumers scored the fresh sample lowest and the
highly oxidized sample highest. This is an interesting result of consumer
behavior. Would a trained assessor, or for that matter a food technologist,
belong to the segment of consumers who like oxidized oil? The answer is
clearly no, because they have been trained and taught to reject off-flavored

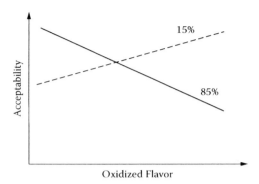

**Figure 2.1** Schematic representation of acceptability versus oxidized flavors for consumers who have a negative slope (85%) and consumers who present a positive slope (15%).

samples. As Table 2.1 indicates, it is a rule that a trained panel doesn't measure acceptability.

What happens when an expert confuses quality control or shelf-life issues with hedonics? Suppose an expert is asked to determine the shelf life of UHT (ultra high temperature processed) milk. The expert tastes the fresh sample and finds it okay, tastes the 2-month sample and finds it okay, and so on. After tasting an 8-month sample the expert rather likes the caramel flavor that has developed during storage; it is sweet and resembles a toffee flavor. Thus the expert states that the sample is okay based on hedonic judgment. What about the consumers? When they are confronted with this 8-month old milk, they find it clearly different from the fresh milk they are accustomed to; and even though the flavor is not objectionable, it is not what they expect in this type of product. Consumers reject the 8-month sample, but the expert had accepted it. Thus we see the dangers of trusting an expert's hedonics.

The final point of Table 2.1 refers to monitoring the performance of assessors. The ISO 8586-1 Standard (1993) covers some aspects of assessor monitoring. Lately (ISO 11132 Standard 2008), a working draft has been developed that, when published, will provide thorough instructions for monitoring performance. Textbooks by Lawless and Heymann (1998), Stone and Sidel (2004), and Meilgaard et al. (2007) recommend assessor monitoring, especially when covering descriptive methods. Researchers (Derndorfer et al. 2005) have published monitoring procedures implemented in freely available software, and institutions such as Matforsk, in Ås, Norway, offer a free package (PanelCheck) for assessor performance monitoring. Thus there seems to be no doubt that different authors and institutions give great importance to this issue. There is a general concern that an assessor does not always perform adequately and that this performance has to be checked. In Chapter 1, I discussed the drawbacks of relying on the judgment of a single expert "life assessor," as proposed by Ellis (1994), because performance can vary from session to session. It has also been demonstrated that it is more probable for older individuals to have sensory impairments than younger individuals (Van Toller 1993). Once an expert has been designated as such, for example, professional wine tasters, there is usually no record of their periodically validating their expertise. Some of these experts are quite aged, and it would be recommended that they go through a set of screening tests to make sure they have no sensory impairments.

## 2.4  General requirements and conditions for sensory tests

### 2.4.1  Testing area

A sensory laboratory for trained panel work should have the following areas:

- A measuring area with individual booths, usually computerized to facilitate data acquisition. Figure 2.2 shows a booth at our laboratory.
- A discussion area fitted with a large table, preferably round, for the panel leader and assessors to evaluate and discuss attributes and descriptors during training for descriptive methodology. Figure 2.3 is a view of our discussion room.
- A sample preparation area that has general cooking and storage facilities.

For possible layouts and other details the ISO 8589 Standard (2007) on design of test rooms for sensory analysis prevents having to guess or invent how to build proper facilities. Having a sophisticated laboratory is a help in conducting sensory tests, but it shouldn't be a limitation. A quiet and pleasant room where assessors can work without interfering with each other should be sufficient. The instruments are the assessors, and no matter how elegant the laboratory or how elaborate the acquisition software, if the assessors have not been adequately screened and trained for the job, the final results will be poor.

Consumers rarely perform their evaluations in sensory laboratories, but rather work in the annex of a cafeteria, a hall of some kind, or in their homes. The same as those for trained assessors, consumer locations should be quiet and pleasant and allow consumers to perform their evaluations without interfering with each other. Possible locations are discussed later when presenting sensory acceptability tests (Section 2.7.3.2).

***Figure 2.2*** Trained assessor measuring the aroma of a beverage in a sensory laboratory equipped with individual booths and a computerized data acquisition system.

***Figure 2.3*** Trained assessors developing texture descriptors for chocolate biscuits.

## 2.4.2   Lighting

General lighting of the laboratory should be with fluorescent lamps. These do not generate too much heat and allow for higher standardization when the goal is daylight-type illumination.

Colored lights are usually an option in sensory laboratories as a means of masking color differences between samples. For example, if an assessor has to decide which of two yogurt samples has a stronger strawberry flavor and one of the samples has a higher red color intensity, it is probable that when in doubt there will be a bias toward the sample with the more intense red color. By placing the samples under a red light, both will look white, and the bias *tends* to disappear—*tends* because often the artifact of using colored lights doesn't work. Either the colored bulbs emit sufficient white light to be ineffective or they only mask the color, but not the intensity.

In shelf-life studies, color masking can be important. For example, minced meat tends to turn brown during refrigerated storage; off-odors also appear during storage. If the researcher considers that off-odor is the critical descriptor, then the researcher will want to mask the color differences so as not to influence assessors. Another example occurred when we were studying shelf life of sunflower oil. During storage, as expected, oxidized flavor developed. What we didn't expect was that the oil kept getting lighter in color, gradually losing its slight yellow tinge. Consumers in Argentina prefer light-colored vegetable oils, and thus there could have been a bias toward preferring stored samples due to appearance. With consumers, we finally decided to test the oils with boiled potatoes as a carrier (Ramírez et al. 2001), and in these conditions the color differences were not detectable.

Alternatives to colored lights in masking color differences are as follows:

- For liquids, cover the glass with aluminum foil and use a straw as shown in Figure 2.4. This method is not entirely recommended because assessors lose the odor of the headspace volatiles while drinking.
- Place the food sample in a colored glass. For example, the standard glass used for sensory evaluation of olive oil is brown (COI Norma T.20/Doc., No. 5, 2007). Oils placed in brown glasses all look alike.
- Color-tinted spectacles have not been mentioned, and there is always the danger of assessors taking their spectacles off momentarily or of getting a glimpse of a sample. Even if this glimpse does not conform to reality ("out of the corner of my eye I think I can see that this sample is darker than the previous one," an assessor may think), the evaluation will be biased.
- Mix the samples with an insipid colorant. This method can be effective but cumbersome.

*Figure 2.4* Trained assessor evaluating a liquid sample.

As mentioned above, consumers rarely perform their evaluations in sensory laboratories. Trying to mask color differences in a central location hall, where lights would need to be changed and windows darkened, is a difficult assignment. Serving samples sequentially can be of help to minimize appearance comparisons among samples, but it does not eliminate the bias of scoring a minced meat sample low in odor acceptability because it is dark brown in color. Consumers can be asked not to consider color when evaluating odor or flavor, but it is doubtful that they will satisfy the request. Consumers receive no sensory training, so it is very probable they don't know how to avoid one attribute influencing another (e.g., color influencing flavor). In certain situations it is difficult to find a good solution to masking color differences in consumer tests.

## 2.4.3   Time of day

There is a general recommendation to conduct sensory tests late in the morning or in mid-afternoon (Poste et al. 1991). However, this is not always possible. Generally, a trained assessor can perform his or her task at any time within reason, that is, not immediately after a hefty lunch that included spicy food and/or alcoholic beverages. With consumers, a general rule would be to avoid tests right after meals and to consider time of day for certain products; an obvious example is not to conduct a consumer test on an alcoholic beverage early in the morning.

## 2.4.4   Carriers

Some food products such as spices need a carrier for sensory evaluation. In research on sensory analysis of mustard (Barbieri et al. 1999), we found that a white sauce made with corn starch and milk was an adequate bland carrier. The ASTM E1871 Standard (2006) on serving protocol for sensory analysis can be consulted for general guidelines, and for some foods specific recommendations are given.

With some products, sample presentation is not the same for a trained panel as for consumers. For example, the ASTM E1346 Standard (2006) on preparing vegetable oils for sensory evaluation indicates that a 10-ml sample at 50°C is to be served to a trained assessor. This would not be adequate for a consumer who would consider any sample of oil tasted on its own at 50°C as awful. As mentioned above, in a study on shelf life of sunflower oil, consumers evaluated the oil samples with boiled potatoes as a carrier (Ramírez et al. 2001).

## 2.4.5   Temperatures of samples

For consumer tests, samples should be served at the temperature they are normally consumed: beer, cold; cheese, room temperature; soup, hot; and so on. Temperatures for trained panel evaluations sometimes differ from normal consumption temperatures. The ASTM E1346 Standard (2006) recommends that vegetable oils be tested at 50°C; although not specifically mentioned, this would include olive oil. The COI Norma T.20/Doc. No. 15 (2007) says that olive oil should be tested at 28°C. For fluid milk the ASTM E1871 Standard (2006) recommends, "serving temperature may be refrigerated or warm depending on test objective," without further details. The IDF 99C Standard (1997) indicates a temperature of 16°C for liquid milk and a temperature of 20°C for reconstituted powdered milk. No documentation can be found to explain this difference in recommended temperatures within the same standard. Documented research needs to be conducted to determine the optimum temperatures for sensory testing of food products by trained panels.

## 2.4.6   Sample size

The ASTM E1871 Standard (2006) gives only a general guideline: "Consider the test objective when determining serving size. Provide enough of the sample to ensure the assessor evaluates the overall product, not just one or two components." We usually provide enough sample for trained assessors to be able to take three sips or three bites; this means approximately 30 g of sample. In consumer tests this quantity is usually doubled.

## 2.4.7  Number of samples

Meilgaard et al. (2007) recommended that for cookies or biscuits, 8 or 10 samples may be the upper limit; with beer, 6 or 8. For products with a high carryover effect, such as hot, spicy foods, only 2 or 3 samples can be evaluated in a session. The number of samples depends on the test. For example, in a paired preference test where a consumer tastes each sample once to decide which she likes best, there is little fatigue, and it is possible to taste 8 beer samples. For a completely hedonic test on beer, with the inclusion of just-about-right scales and acceptability judgments on different attributes, each beer sample might have to be tasted 4 times. If 8 samples were to be included, this would mean 4 times × 8 samples = 32 tastings of beer; clearly too much. It must also be considered that sensory assessors tend to lose their motivation before losing their ability; and once they have lost motivation, their ability will decrease.

## 2.4.8  Coding and order of presentation

General practice is to code samples with three-digit numbers chosen at random. Depending on the test, order of presentation can be balanced or random. In consumer tests the general practice is to balance order of presentation and first-order, carry-over effects (MacFie et al. 1989). For discrimination tests, such as the triangle or paired comparison tests, order of presentation is also balanced (Meilgaard et al. 2007). For descriptive tests, where there is no evidence that order of presentation has an effect on evaluations, randomized order of presentation for each assessor is recommended.

## 2.4.9  Palate cleansers

To avoid saturation, trained assessors usually use a palate cleanser before evaluation and between samples. Different types of products require different palate cleansers. The ASTM E1871 Standard (2006) recommends warm water for products that leave an oily residue, and milk or cream cheese for products with garlic or spices. For certain products only extended periods of rest will sufficiently eliminate the carryover effects of a product. In our laboratory we use the following palate cleansers: water, saltless crackers, bread, and Granny Smith apples. During the initial training periods with a new product category, these palate cleansers are offered to assessors who reach a consensus on which one they consider most effective. If crackers, bread, or apples are chosen, they are accompanied by water. Usually only water is used for consumer tests; we have found that the consumers' task is sufficiently complicated by the samples and scales while expecting them to remember to eat a piece of apple and rinse with water between samples. Even if they are given this instruction, not all of them comply with it.

## 2.5   Physiological factors

Trained sensory assessors should not participate, or should at least warn the panel leader, when they suffer from any of the following:

- Colds or any other ailments such as allergies that lead to nose congestion.
- Skin allergies that may affect their sense of touch.
- Dental problems that may limit or alter texture evaluations.
- Emotional stress that can lessen concentration on the sensory evaluation task. Sometimes it is the other way around: a sensory session can serve as a helpful distraction.
- Time pressures that can lead the assessor to rush through a session.

Trained assessors are taught that during the half hour previous to a sensory session, they should not smoke, drink coffee (or *mate* in Argentina), or consume any other product that may affect their senses. However, if they inadvertently indulge, they should warn the panel leader of their mishap.

Table 2.2 presents physiological factors that can influence sensory evaluations. Although this table presents examples, it is often difficult to anticipate which of these effects might be present in other situations. Palate cleansers and sufficient rest periods between samples can be used to avoid the errors that can occur due to these physiological factors.

## 2.6   Psychological factors

### 2.6.1   Expectation error

Trained assessors and consumers usually find what they expect to find. In an SSL study of fluid milk, when assessors are called in every 2–4 days to evaluate aged flavors such as sour, fermented, or oxidized, they soon catch on that they are involved in a shelf-life study. This leads them to expect the presence of aged flavors, and once found they are likely to repeatedly find them for subsequent storage times. Blind control samples can be introduced to purposefully mislead assessors; but not a single blind control in each session, as this is a pattern soon learned by assessors. We suggest zero to two blind controls randomly allotted to sessions. By definition consumers are more naïve than trained assessors, but they also start speculating as to why they are called in every 2–4 days (to continue with the milk example) to express their acceptance/rejection for a set of

*Table 2.2* Physiological Factors That Influence Sensory Verdicts

| Factors | Description | Examples |
|---|---|---|
| Adaptation | A decrease in or change in sensitivity to a given stimulus as a result of continued exposure to that stimulus or a similar one | A respondent, having tested a set of soft drinks, is unable to properly score the sweetness in a less sweet sample |
| Cross-adaptation | Adaptation caused by previous exposure to a different stimulus | Insensitivity to the sweetness of sugar caused by previous exposure to another sweetener |
| Cross-potentiation | Positive adaptation (or facilitation) caused by previous exposure to a substance of a different flavor | An observer having been exposed to a sour candy perceives more sweetness in a subsequent sweet candy |
| Enhancement | Effect of presence of one substance increasing the perceived intensity of another presented simultaneously | Amyl alcohols enhance the perception of the rose flavor of phenylethanol |
| Synergy | Effect of the simultaneous presence of two or more substances increasing the perceived intensity of the mixture above the sum of the intensities of the components | The impact on taste of monosodium glutamate is enhanced in mixtures with 5′ ribonucleotides |
| Suppression, masking | Effect of the presence of one substance decreasing the perceived intensity of another presented simultaneously | Sweetness often suppresses or masks bitter or sour tastes. "A spoonful of sugar helps the medicine go down." |

*Note:* Chambers IV, E., and M. Baker Wolf. 1996. ASTM Manual Series MNL 26: Sensory testing methods. West Conshohocken: American Society for Testing Materials.

milk samples. Once they start rejecting the samples, they will expect the following samples to be worse and reject them also. The reverse storage design of shelf-life studies described in the following chapter (Section 3.3.6.2) eliminates this type of expectation error.

## 2.6.2   Error of habituation

This refers to the tendency to repeat the same response when samples show low levels of variation. This can occur in SSL studies where differences between samples from contiguous storage times can be small. To avoid this error, blind controls or blind doctored-up samples are presented randomly at measuring sessions. If these special samples differ widely as to what can be expected, assessors can think that they only have to report a difference when it is as wide as the special sample; thus, these samples have to be carefully designed.

### 2.6.3 Stimulus error

The trained assessor or consumer can be influenced by irrelevant information. In a study on commercial cracker-type biscuits (Martínez et al. 2002), we found that brands were engraved on some of the samples, and we could not adequately hide this information from consumers. We asked them not to consider the brand in their evaluation, but a stimulus error could have occurred.

### 2.6.4 Logical error

This error occurs when two or more characteristics of a product are associated in the mind of an assessor. Assessors more prone to committing this error are those who have some knowledge of the product's transformations. For example, when evaluating stored UHT milk, an assessor is likely to score a dark-colored sample high in caramel flavor; the assessor knows that the non-enzymatic browning reaction generates caramel-type flavors.

### 2.6.5 Halo effect

When more than one attribute of a sample is evaluated, the ratings of the different attributes tend to be influenced by one another. The halo effect is generally associated with consumer studies and can be difficult to eliminate. If an orange juice develops a dark, unnatural color during storage and a consumer is asked to score his/her acceptability of the juice's appearance, the score might be a 4 on a 1–10 hedonic scale. If then asked to score its flavor, no matter how good the juice tastes, it is improbable that the flavor score will be higher than 5 or 6. One possibility is to have consumers evaluate the flavor of the samples under colored lighting, and then change the lights and ask them to evaluate the appearance. The problem with this approach is that consumers are likely to be so distracted by the artificial setting that their scores won't mean much.

### 2.6.6 Positional bias

In consumer studies there is an order-of-presentation effect (Hottenstein et al. 2008). In triangle tests, the odd sample is detected more often if it is in the middle position (Meilgaard et al. 2007). To avoid these positional biases, refer to Section 2.4.8, where order of presentation criteria was discussed.

*Figure 2.5* Biscuit samples at different storage times.

## 2.6.7   Contrast effect and convergence error

Consider a study on stored biscuits where there are four samples (Figure 2.5):

- Sample A: fresh and highly acceptable
- Sample B: different but close to A due to relatively short storage time
- Sample C: midway between A and B
- Sample D: aged and highly objectionable

If a consumer evaluates Sample A first and then Sample C, it is probable that Sample C will be rejected not on its own value but in contrast to A. Similarly, if Sample D is evaluated first and then Sample C, it is probable that Sample C will be accepted not on its own value but in contrast to D.

When using a hedonic scale, if Sample A is evaluated first, Sample D second, and Sample B third, it is likely that Sample B will receive a similar hedonic score to Sample A. Due to the presence of D, scores for Samples A and B, which are similar to each other, tend to converge.

To avoid these contrast and convergence effects, refer to Section 2.4.8, where order of presentation criteria was discussed.

## 2.6.8   Mutual suggestion

Reactions of other members of the panel can influence the response of a trained assessor. The common sense suggestion is to have individual booths so that no interaction between assessors takes place. When studying the shelf life of a product that developed oxidized flavor, I was one of the assessors. I had evaluated three of the five samples we received that day and had found none with an oxidized flavor. When I was about to start on the fourth sample, I heard one of my fellow assessors make an involuntary sound of disgust. She had obviously tasted a highly oxidized flavor. This "suggestion" put me in a state where expectation error was highly probable.

## 2.6.9   Lack of motivation

Previously we mentioned that assessors lose their motivation before losing their ability. Sometimes their lack of motivation is verbalized, and other times it is reflected in poor or unexpected results. In certain SSL studies where assessors are asked to repeatedly evaluate samples with increasing levels of deterioration over a period of months, lack of motivation can be an issue. In an SSL study of powdered milk, we had the case of an assessor who didn't say anything; but when the study was advanced and samples had developed a definite oxidized flavor, he started scoring all samples with negligible values for this flavor. A conversation with this assessor revealed that he was fed up with tasting these oxidized samples of powdered milk, and I suppose we can't really blame him!

In the storage of food products, other issues apart from SSL estimation are sometimes of interest. For example, a project leader may want to correlate oxidized flavor developed during storage of powdered milk with a chemical index, thus prolonging the study beyond the actual SSL to obtain this correlation. Tasting small portions of these oxidized samples may not produce physical damage, but it is hardly a pleasant experience. Is it ethical, in the pursuit of scientific knowledge, to ask people to go through unpleasant sensory experiences? In our institute we have an ethics committee that deals with these issues. This committee is formed by the project leader, a trained sensory assessor involved in the project, and a research colleague from another department with food chemistry and microbiology knowledge. Common sense and "play it safe" have always been the rule in the committee's decisions.

## 2.6.10   Capriciousness versus timidity

Some assessors tend to use the extremes of the sensory scale to rate small sensory differences, and others tend to use small portions of the scale. For trained panels these problems can be solved by using appropriate standards. For consumer studies the issue of capriciousness versus timidity would depend on the type of scale used. In Chapter 4, survival analysis methodology will be presented to estimate SSL based on consumers' answering if they accept or reject samples with different storage times. Capriciousness could work both ways; that is, a consumer could be capricious and thus decide to accept all or most samples, or be capricious and decide to reject all or most samples. Timidity would work likewise. If consumers are asked to use a hedonic scale, then a capricious consumer would tend to separate two samples using extremes of the hedonic scale, and a timid consumer would tend to use a limited portion of the scale. The issue could be resolved if each consumer's scores were standardized by subtracting the consumer's mean and dividing by standard deviation. This practice is not usually done when analyzing the data by analysis of variance, although it has been recommended when statistical tests such as cluster analysis are performed (MacFie 2007).

*Table 2.3* Questions Answered by Sensory Evaluation and Methods Used

| Question | Methods |
| --- | --- |
| Are there sensory differences between the samples? | Discrimination tests |
| In what way are the samples different? | Descriptive analysis |
| Do consumers care about the differences? | Affective tests |

## 2.7   Sensory evaluation methods

Table 2.3 shows the basic questions that we want answered by sensory evaluation and the principal methods used to answer these questions. In-depth description and discussion of sensory methods can be found in the following textbooks: Lawless and Heymann (1998); Stone and Sidel (2004); Meilgaard et al. (2007). Following is a summary description of sensory methods with some considerations of the application of these methods to SSL studies.

### 2.7.1   Discrimination tests

Discrimination tests can be classified in the following two categories:

1. Overall difference tests: these are designed to answer whether there is any sensory difference between the samples. The most widely used test in this category is the triangle test. Another overall difference test of interest to SSL is the difference from control test.
2. Descriptor difference tests: these are used to find differences in a specific descriptor. For example: "rank these three samples by sweetness" and all other attributes are ignored. The paired comparison and ranking tests are usually used when the descriptor to be found is known beforehand by the researcher.

In SSL tests, the critical descriptor may be unknown. This can occur if there is insufficient experience with the product or if no preliminary tests have been performed to determine the critical descriptor. In a study of the SSL of non-fat yogurt, the critical descriptor was hard to define among the following attributes: presence of whey when opening the pot, acid taste, loss of sweetness, and cheesy flavor. In other SSL tests, the critical descriptor is known; for example, during oil storage the development of oxidized flavor was the critical descriptor (Ramírez et al. 2001).

#### 2.7.1.1   Triangle test

Each assessor receives three coded samples: Two samples are identical, and one is different from the other two. The assessor's task is to pick out the odd sample. Complete details on how to conduct this test can be found in the ISO 4120 Standard (2004).

Generally speaking, discrimination tests are not considered appropriate for quality control or SSL studies. For example, in an SSL study of a packaged biscuit, a comparison between freshly made biscuits and biscuits stored for 6 months could be of interest. If a triangle test were to be performed, in all probability significant differences would be found, though this does not mean the product has reached the end of its shelf life. A consumer will probably find the 6-month-old biscuit okay even though it has lost a certain degree of crispiness. Chapters 4 and 6 will present methodology to estimate SSL when a sensory difference between the fresh product and one at the end of its shelf life is tolerated by consumers. This would be the general case; however, there are products and manufacturers who are very strict and do not want consumers to detect any sensory difference between the fresh and aged products. In this case a triangle test would be appropriate for determining SSL.

### 2.7.1.2   Example of sensory shelf life (SSL) determined by a triangle test

A manufacturer of a traditional herb-based alcoholic beverage wants to establish the SSL of the product with the criterion that no sensory difference be detected between the fresh and aged product.

The objective is to compare the fresh product versus samples stored at 25°C (considered medium to high ambient temperature for the region where the product is sold) for 4, 6, and 8 months. The product is a complex mixture, and thus critical sensory changes cannot be limited to a single descriptor. Based on the above classification of discrimination methods, a triangle test is chosen for comparison purposes at each storage time. To determine the number of assessors, the first step is to analyze the null and alternative hypothesis. As an illustration we shall consider *fresh* (F) versus *stored for 6 months* (S6):

Null hypothesis—H0: F = S6
Alternative hypothesis—H1: F ≠ S6

If H0 is rejected when it should have been accepted, we would be committing Type I error, and to protect against this error we would choose a low value of $\alpha$. If H0 is accepted when it should have been rejected, we would be committing Type II error, and to protect against this error we should choose a low value of $\beta$. What are the consequences of committing these errors?

- Type I error: if we found a difference at 6 months' storage and there was no difference at 4 months' storage, we would establish the SSL at

4 months. Now if this had been an error, that is, in fact there was no difference at 6 months, we would be stamping 4 months shelf life on the product when we could have been stamping 6 months. Obviously there would be no consumer complaints, but the company would have additional distribution and rotation costs. This Type I error can be considered of relatively low risk, thus an $\alpha = 10\%$ is chosen.

• Type II error: if no difference was found after 6 months of storage and there was a difference at 8 months, we would establish the SSL at 6 months. Now if this had been an error, that is, in fact there was a difference at 6 months, we would be stamping 6 months shelf life on the product when we should have been stamping 4 months. Consumers drinking the beverage after 5–6 months of storage would probably be complaining. Even if they don't complain, the brand image would suffer and sales would diminish. It is considered that not too many consumers would be drinking the product close to the end of its shelf life, thus the risk is average. It is established to work with a 95% certainty (i.e., $\beta = 5\%$) that no more than 30% of the population (i.e., $p_d = 30\%$) can detect a difference. The ISO 4120 Standard (2004) states that $p_d$ values between 25% and 35% can be considered as medium.

With $\alpha = 10\%$, $\beta = 5\%$, and $p_d = 30\%$; Table A.3 of the ISO 4120 Standard (2004) recommends 54 assessors for each triangle test.

A company uses a buffering system in its production line that aids in maintaining a product with constant quality. This consists of having each production batch go through quality control procedures, and once approved the batch is transferred to a huge tank that holds the equivalent of 30 batches. Any variation in the sensory quality of one batch is thus diluted. Bottling of the beverage is done from the huge tank. How many 1 liter bottles need to be taken from this tank for the SSL test:

1. Each triangle: 54 assessors × 3 glasses × 30 ml = 4860 ml.
2. Total of three triangle tests (fresh versus 4 months, 6 months and 8 months): 4860 × 3 = 14580 ml.
3. The beverage has a high alcohol content, so it is diluted 1 part beverage + 1 part water; thus the total volume of beverage that is necessary: 14580 ÷ 2 = 7290 ml.
4. A few bottles, more or less, have a minor influence on the cost of the project, so to be on the safe side, 18 bottles filled consecutively are taken from the production line. One of the worst things that could happen would be for a bottle to break and thus ruin an 8-month storage experiment. Taking 18 bottles would even allow repeating a triangle test, should a mistake be suspected in a particular test.

It has already been established that the product suffers no sensory changes when stored at 2°C in the dark for 1 year. The storage procedure will consist of a reverse storage design which will be presented formally in Chapter 3. This will consist of the following:

- Time = 0: place 15 bottles in the refrigerator at 2°C and three bottles in the constant temperature chamber at 25°C
- Time = 2 months: remove three bottles from the refrigerator and place them in the 25°C chamber.
- Time = 4 months: remove three bottles from the refrigerator and place them in the 25°C chamber.
- Time = 8 months: remove all nine bottles from the 25°C chamber and place them in the refrigerator.

After 8 months in the refrigerator there will be: nine bottles of *fresh* sample (bottles that remained at 2°C for 8 months); and three bottles stored at 25°C for each of the following times: 4 months, 6 months, and 8 months. It was established that the product could be stored at 2°C for at least 1 year without sensory changes, but this refers to an initially fresh sample. A sample that has been stored for 8 months at 25°C has already initiated deterioration, and when placed at 2°C the deterioration will slow down but cannot be neglected. Thus when the 8-month storage period has been completed, the triangle tests should be conducted within the following week.

The design of the triangle tests is fully balanced, that is:

- As this alcoholic beverage is highly saturating, each assessor performs a single triangle test per session. Each session is conducted on a separate day. A third of the assessors receive the triangle tests in the following order: F versus S4, F versus S6, and F versus S8. Another third receives: F versus S6, F versus S8, and F versus S4. And the other third receives: F versus S8, F versus S4, and F versus S6. Allocation of serving orders to assessors should be at random.
- Within each triangle test the standard balanced order of presentation is respected. For example, for F versus S4, there are six possible presentation orders: F-F-S4, F-S4-F, S4-F-F, S4-S4-F, S4-F-S4, and F-S4-S4.

The first choice to perform the triangle tests would be trained assessors. Not very many laboratories have 54 trained assessors they can count on. An alternative is to recruit regular consumers of the herb-based beverage. Who can be considered a *regular consumer*? This beverage is consumed mostly on weekends at bars or as a drink at home. It is decided to recruit men or women between the ages of 20 and 50 years who have consumed the beverage in the last two weeks. Previous to the first triangle

*Table 2.4* Results of the Triangle Tests between Fresh and Stored Samples
of an Herb-Based Alcoholic Beverage

| Triangle test | Total number of responses | Number of correct answers | % Significance level | Upper 95% confidence limit |
|---|---|---|---|---|
| Fresh vs. 4 months | 54 | 22 | 15.6 | 28 |
| Fresh vs. 6 months | 54 | 24 | 5.8 | 33 |
| Fresh vs. 8 months | 54 | 30 | 0.1 | 50 |

test, consumers are instructed on the procedure of the test. To aid com-
prehension, a dummy triangle test is suggested of water versus a 1.6%
sucrose solution. It is quite easy to discriminate between these two sub-
stances, so all assessors should get the triangle right. All assessors indicate which was their odd sample, so that the whole group realizes that the
odd sample can be in any position and also that the odd sample can be the
weak or strong stimuli.

As mentioned above, the beverage is diluted: 1 part beverage + 1 part
water. The diluted beverage will be referred to as the sample. Thirty ml of
each sample is served at room temperature, in 70-ml glasses and under white
daylight-type illumination. Individual booths can be used but are not absolutely necessary. An individual table for each assessor should be adequate.

The number of correct responses in each one of the triangle tests is
in Table 2.4. The significance levels were calculated using the following
Excel®- function:

```
=  (1-BINOMDIST(B4-1,A4,0.3333,TRUE))*100;
```

where A4 holds the total number of responses and B4 the number of correct responses.

The upper limit of the confidence interval for percent of the population of assessors who can discriminate between samples was calculated
using the following expression (ISO 4120 Standard 2004):

$$\text{UL\%} = \left( \left( 1.5 \left( \frac{x}{n} \right) - 0.5 \right) + 1.5 z_\beta \sqrt{\frac{\left( nx - x^2 \right)}{n^3}} \right) \times 100$$

where:
  UL% = upper 95% confidence limit
  $n$     = total number of responses
  $x$     = number of correct responses
  $z_\beta$     = one-tailed normal distribution coordinate corresponding to
        the adopted β value (1.64 for β = 5%)

For 6 and 8 months' storage, the significance levels were below the adopted value of $\alpha = 10\%$; and the upper limits were above $p_d = 30\%$ adopted in the design. Thus there was evidence that for these storage times, samples were different from the fresh sample. For 4 months' storage, the significance level was 15.6%, above $\alpha = 10\%$; and the upper limit was 28%, below $p_d = 30\%$. Thus it was concluded that the SSL of the herb-based beverage was 4 months. Observing the results from Table 2.4 would probably justify repeating the experiment for samples stored 4, 5, and 6 months to see if shelf life could be extended to 5 months.

### 2.7.1.3 Paired comparison test

Each assessor receives two coded samples. The assessor's task is to choose the sample that has the highest intensity of a pre-defined descriptor. Complete details on how to conduct this test can be found in the ISO 5495 Standard (2005). This test is also known as the 2-AFC test (2 sample, alternate forced choice test). It has been shown (O'Mahony and Rousseau 2002) that the paired comparison test is more powerful in finding differences than the triangle test. Of course, to use the paired comparison test, the descriptor to be found by the assessors must be known beforehand. In some SSL studies this can be the case. For example, in a study on the SSL of *dulce de leche* (Garitta et al. 2004), the critical descriptor was plastic flavor. This flavor was perceived by assessors when tasting samples obtained from the borders of the dulce de leche pots in contact with the plastic polystyrene containers. So if the SSL criterion was "absence of plastic flavor," an experiment similar to the one described in the previous section (2.7.1.2) could be designed, but using the paired comparison test instead of the triangle test. A caveat of the paired comparison test: Assessors have to be trained in the descriptor they are told to find. It was shown in Section 2.7.1.2 that for $\alpha = 10\%$, $\beta = 5\%$, and $p_d = 30\%$, 54 assessors are recommended. Using these same parameters for the paired comparison test (ISO Standard 5495 2005) entails using 96 assessors. For the dulce de leche example, all these assessors would have to be trained in searching for plastic flavor. The number of assessors can be reduced by having each assessor test more than one pair, but this procedure requires careful statistical design and analysis (Meyners and Brockhoff 2003).

### 2.7.1.4 Difference from control test

Forced-choice difference tests like the triangle or paired comparison tests are more discriminating than using scales to measure differences (Kim et al. 2006). However, there are situations where it is more important to know the size of the difference than the existence of a difference. This is particularly true for SSL studies. In Chapter 6 the cut-off point methodology will be presented, which is based on the size of the difference in acceptability perceived by consumers and the corresponding size perceived by a panel of trained assessors.

To illustrate the difference from control test, consider an experiment where it is of interest to determine sensory differences in flavor between powdered milk stored for 8 and 12 months at 20°C in comparison to a fresh product. The samples will be measured by a panel of 20 assessors who regularly perform discrimination and descriptive tests, not necessarily with powdered milk. How much powdered milk needs to be stored? Powdered milk is consumed reconstituted in a concentration of 130 grams/liter. For the final sensory test, the following volume will be required:

30 ml × 20 assessors × 4 samples (control, blind control, 8 months, and 12 months) = 2400 ml.

The smallest packet is 500 grams. One packet for the controls, and one each for the 8-month and 12-month storage times would be sufficient; to be cautious, two packets for each condition will be used: that is a total of six 500-gram packets. The packets are chosen from a production batch that had passed quality control procedures.

Powdered milk suffers no change when frozen. The storage procedure will consist of a reverse storage design, which will be presented formally in Chapter 3. This will consist of the following:

- Time = 0: place four packets in the freezer and two packets in the 20°C chamber
- Time = 4 months: remove two packets from the freezer and place them in the 20°C chamber
- Time = 12 months: remove all four packets from the 20°C chamber and place them in the freezer

After 12 months, in the freezer there will be two packets of *fresh* sample (packets that remained in the freezer for 12 months), and two packets stored at 20°C for each of the following times: 8 months and 12 months.

Thirty ml of each sample is served at room temperature, in 70-ml glasses and under white daylight-type illumination. Individual booths can be used but are not absolutely necessary. An individual table for each assessor should be adequate. Each assessor receives a tray with four samples; one of these samples is coded with "C" and the others with three-digit codes. The coded samples are presented to each assessor in random order. The score sheet is shown in Figure 2.6.

Results are shown in Table 2.5. When samples are measured at different time points, like those in the present example, it is recommended that

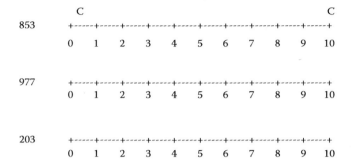

Assessor: ..........................                          Date: 14/01/09

First you must taste the sample labeled as C. After this taste each of the numbered
samples and compare their flavor to C. You can re-taste C if you want to. One or
more of the numbered samples can be equal to C.

*Figure 2.6* Score sheet used for difference from control test.

repeated-measure procedures be used to analyze the data, especially as
time is a variable that usually cannot be randomized (Mead 1988). In the
present case, as a reversed storage design was used, the time factor could
be randomized, and thus the experiment can be analyzed like a randomized block experiment considering assessors as blocks. This analysis of
variance can be done with Excel® using the option of two-factor analysis
of variance with only one sample per group. Another option is to use one
of the many statistical packages available; we use Genstat 11th Edition
(VSN International, Hemel Hempstead, United Kingdom). For the present
analysis the following model is set up in Genstat:

```
BLOCKSTRUCTURE Assessor
TREATMENTSTRUCTURE Sample
ANOVA [PRINT=anova,mean;FPROB=yes;PSE=lsd]Dif
```

The output is shown in Figure 2.7. From this output the conclusion is
that up to 8 months' storage there is no significant change with respect to
the fresh sample. From 8 months on deterioration accelerates, and this is
shown by the significant difference found for 12 months' storage. As will
be discussed in Chapter 4, an important issue in defining the shelf life
of the product is to determine the probability of a consumer's rejecting
samples as a function of storage time. Trained assessors found the sample with 12 months' storage significantly different from the fresh control
sample, but this could have been due to the development of a more pronounced creamy color that consumers do not object to or to an oxidized

*Table 2.5* Results of a Difference from Control Test
Performed by 20 Assessors Who Measured Three Samples
(blind control, 8 months' storage, and 12 months' storage)

| Assessor | Blind control | 8 months | 12 months |
|----------|---------------|----------|-----------|
| 1 | 1 | 0 | 5 |
| 2 | 2 | 2 | 6 |
| 3 | 0 | 3 | 4 |
| 4 | 3 | 1 | 3 |
| 5 | 4 | 4 | 6 |
| 6 | 2 | 1 | 6 |
| 7 | 1 | 2 | 3 |
| 8 | 4 | 5 | 4 |
| 9 | 0 | 1 | 5 |
| 10 | 0 | 0 | 3 |
| 11 | 1 | 2 | 5 |
| 12 | 1 | 0 | 4 |
| 13 | 3 | 4 | 3 |
| 14 | 2 | 2 | 5 |
| 15 | 4 | 3 | 3 |
| 16 | 1 | 3 | 4 |
| 17 | 1 | 3 | 5 |
| 18 | 2 | 4 | 6 |
| 19 | 2 | 2 | 4 |
| 20 | 1 | 0 | 7 |

flavor that consumers find highly objectionable. In SSL the consumers
have the last word.

## 2.7.2 Descriptive tests

Descriptive tests are used to answer the question "In what ways do the
samples differ?" (Table 2.3). Results from a descriptive test encompass a
complete description of the sensory characteristics of a product that is
useful for:

- Providing the basis to understand sensory acceptability
- Investigating how formulation or process changes affect sensory
  properties
- Correlating instrumental measurements such as chromatography
  results to sensory data
- Establishing the critical descriptors important to quality control or
  shelf-life studies
- Guiding the research and development of food products

```
Analysis of variance

Variate: Dif

Source of variation  d.f.     s.s.    m.s.    v.r.  F pr.

Assessor stratum        19   45.600   2.400   1.55

Assessor.*Units* stratum
Sample                   2   93.100  46.550  30.03  <.001
Residual                38   58.900   1.550

Total                   59  197.600
```

**Tables of means**

```
Variate: Dif
Grand mean  2.80
   Sample   Control    S8     S12
            1.75      2.10   4.55
```

**Least significant differences of means (5% level)**

```
Table        Sample
rep.             20
d.f.             38
l.s.d.        0.797
```

*Figure 2.7* Analysis of variance of difference from control test on powdered milk.

The ISO 13299 Standard (2003) classifies profiling techniques as follows:

- Conventional profiling: Assessors, seated in booths, score each sample on a preselected set of attributes and scales.
- Consensus profiling: Through consensus discussion, the panel, seated around a table, develops its own terminology and scores pertaining to the sample set presented.
- Free-choice profiling: Assessors, seated in booths, are free to choose their own terminology and scale. A profile is derived statistically, for example, via generalized Procrustes analysis.
- Time-intensity profiling: Assessors, seated in booths, record the intensity of an attribute over time.

Conventional profiling is the most-used technique, and there are variations related mainly to descriptor development and assessor training. In Quantitative descriptive analysis (QDA) (Stone and Sidel 2004), the recommended number of judges is between 10 and 12, and during the training period the panel leader has less influence than in other methods in defining descriptors. Sensory spectrum (Meilgaard et al. 2007) is a variation of QDA. The assessors do not generate a panel-specific vocabulary; the language used to describe particular products is chosen *a priori* and remains the same for all products within a category over time. Additionally, the scales are standardized and anchored with multiple reference points, which usually extends the training period in comparison to QDA. Once the training is completed, the samples' profiles are defined in a similar manner to QDA. As pointed out by Lawless and Heymann (1998), QDA and sensory spectrum techniques have been adapted in many different ways. Academic researchers frequently employ the general guidelines of these methodologies to evaluate products.

When performing descriptive analysis it is recommended to cover all attributes of the product. From an assessor's point of view it is not appropriate to ask him or her to exclude certain attributes. For example, when evaluating a caramel sweet, it may be of interest to a certain project to only measure taste. If odor is excluded, when an assessor finds a sample with a higher vanilla note, he or she will probably dump the difference in vanilla on the sweetness scale; thus, it would be best to include the vanilla descriptor, even though it is of no direct interest. Another issue with excluding attributes is that preconceived ideas may prove to be erroneous. For example, in a study of the SSL of a cracker, it could be supposed that the critical descriptor is the loss of crispness. However, due to high-quality packaging, moisture uptake during storage is negligible and the critical descriptor could be a fishy flavor due to fat oxidation.

Another recommendation is to profile more than one sample. Assessors are good at comparing sensory properties among samples, but not good at providing absolute values. In SSL studies the fresh control sample should always be present when possible. In some instances descriptor scales are anchored at the center with the control sample. Figure 2.8 shows two scales corresponding to the measurement of a stored sunflower oil sample. A

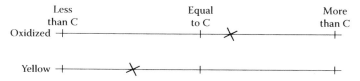

*Figure 2.8* Descriptive scales anchored with the control sample.

stored sample should always have a higher oxidized flavor than the fresh control sample, so the lower part of the scale would be unnecessary. But as ratings of other descriptors, such as yellow color, diminish in stored samples, it is preferable to use a single-scale format for all descriptors.

The number of recommended assessors for descriptive tests is 8–12. For SSL tests that have to be followed for relatively long periods, keeping all panel members throughout the study can be difficult. Usually the number of descriptors followed during the study is limited to those that are critical, in the range of three to six, depending on the product. This small number of descriptors and the continuous exposure to the product helps in calibrating the panel, and this compensates for the loss of one or two assessors along the way.

For descriptive analysis assessors are screened and trained. The ISO 8586-1 Standard (1993) presents adequate guidelines for screening and training assessors for descriptive analysis. A desirable quality in a trained assessor is to associate and verbalize sensory perceptions. This is important in the first stages of a test where the descriptors are developed. In SSL studies, descriptors not usually associated with the product may arise. For example, in a shelf-life study on powdered soups, a hard-to-define flavor appeared in the stored samples. The breakthrough was obtained when an assessor associated this flavor with the smell she experienced when passing by a meat flour plant on the outskirts of the city. Panel leaders have commented that assessors who don't say much during the descriptor development stage sometimes perform just as well when measuring as assessors who are good at verbalizing. However, it must be remembered that if it weren't for these last assessors, the quiet ones would have nothing to measure.

ISO 13299 Standard (2003) suggests three procedures for choosing optimal descriptors:

1. Use existing terminology and reference standards.
2. Use the panel in special sessions to develop the terminology it will use; the final set of descriptors is defined by the panel and the panel leader with the help of reference samples.
3. Use the panel in special sessions to develop the terminology it will use; the final set of descriptors is obtained following the procedure defined by ISO 11035 Standard (1994). This standard describes a recommended method of identifying and selecting discriminating terms using a set of prepared training samples; the final number of terms is obtained by stepwise elimination using statistical techniques.

In SSL studies it is customary to use existing terminology and reference standards. For example, for powdered milk, oxidized flavor should be monitored, and there are published procedures for preparing an adequate reference (Hough et al. 1992). It would not be necessary for the panel to work on developing this obvious descriptor.

Stone and Sidel (2004) proposed the use of an unstructured scale for the QDA method of the type shown in Figure 2.8. For the sensory spectrum method, Meilgaard et al. (2007) indicated that assessors write (or enter) the actual number for the intensity of each attribute using a 0–15 scale measured in tenths. Quoting Lawless and Heymann (1998, p. 245): "Most published studies have found about equal sensitivity to the different scaling methods, provided that the methods are applied in a reasonable manner." In our laboratory we prefer using structured scales of the type shown in Figure 2.6. In the initial stages of a descriptive analysis—when the panel is searching for references and reaching a consensus on their values on the scale—having structured scales simplifies the discussion as assessors can easily call out their scores during discussions. For the measuring stage assessors introduce their scores on a computer screen as shown in Figure 2.9.

In descriptive analysis it is recommended to have assessors measure the products on more than one occasion (Stone and Sidel 2004). This allows for controlling the performance of the panel as a whole and each assessor individually. For example, if we want to check whether a particular

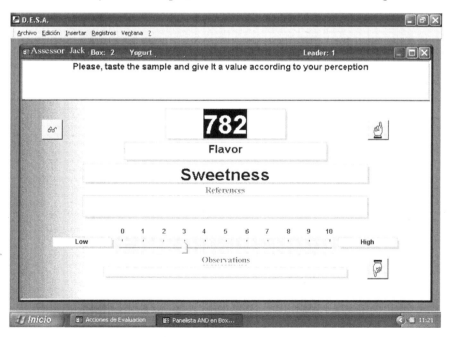

*Figure 2.9* Screen visualized by a trained assessor to measure sweetness in a yogurt sample.

assessor is discriminating between the samples, an analysis of variance on his or her individual scores can be performed, but repetitions of scores on the same sample are necessary. Another reason for conducting repeated tests on the same samples is to increase the discriminating power of the test. Stone and Sidel (2004, p. 222) recommend "about four trials from each subject for each product." In our sensory laboratory we follow the general rule of three trials. In SSL tests it is not always possible to comply with so many replicates due to limited storage space during the shelf-life study. It can also be fatiguing for assessors to evaluate a large number of samples with unpleasant storage flavors. In these cases we have reduced replications to two.

Analysis of variance (ANOVA) is the most widely used tool for analyzing data from descriptive analysis. The factors of variation to be considered in the ANOVA model to be applied are session, replication, order of presentation, assessor, and specific treatment factors. A good starting point in understanding ANOVA models for sensory analysis is O'Mahony's book (1986). A more comprehensive treatment of the subject is given by Lea et al. (1997).

Of particular interest to the application of ANOVA to SSL studies is the consideration of repeated measures designs. These designs occur typically in biological studies of different sorts, where different treatments are applied to different subjects, and their response is followed over time. For example, if the effectiveness of a ration supplement is to be tested on pigs, a group of pigs is randomly divided into two lots. One lot receives the control ration and the other lot receives the supplemented ration. The weight of each pig is registered at specific periods. By the end of the experiment, there is an interest in knowing whether the ration supplement affected the pigs' weight and also if this effect changed over time. In SSL studies a repeated measures design could be as follows. Suppose there is an interest in testing the effect of the addition of an iron complex on the SSL of pasteurized milk. The milk is to be stored at 5°C for a total of 25 days, and sampled every 5 days. On each sampling day a trained sensory panel will measure the oxidized flavor of the control sample and the sample with iron. In this case the subjects are the trained assessors, and the treatments would be iron addition and storage time. Mead (1988) described repeated measures design and analysis in detail.

Results from descriptive analysis are presented in tabular and graphical format. Table 2.6 shows results from a descriptive analysis performed in our laboratory on eight commercial samples of beer. It should be noted that each number in the table is the result of averaging 30 numbers: 10 assessors × triplicate. It is customary to include letters next to average numbers in such a table, indicating which samples differ significantly in relation to the least significant difference (LSD). We prefer to include the LSD value to avoid the cluttering of tables with myriad little letters.

Table 2.6 Results of a Descriptive Analysis of Eight Commercial Samples of Beer

| Samples | Acetone | Acetaldehyde | Sulfurous | Diacetyl | Bitter | Acid | Sweet | Light | Oxidized | Color |
|---|---|---|---|---|---|---|---|---|---|---|
| A | 20 | 1 | 37 | 7 | 42 | 19 | 10 | 9 | 19 | 36 |
| B | 20 | 1 | 16 | 7 | 28 | 10 | 12 | 21 | 14 | 15 |
| C | 26 | 0 | 30 | 10 | 61 | 16 | 2 | 1 | 27 | 56 |
| D | 29 | 19 | 21 | 9 | 37 | 23 | 15 | 5 | 21 | 87 |
| E | 30 | 1 | 33 | 6 | 54 | 28 | 7 | 6 | 31 | 33 |
| F | 20 | 1 | 14 | 8 | 36 | 14 | 9 | 5 | 22 | 40 |
| G | 8 | 2 | 11 | 7 | 35 | 14 | 9 | 18 | 23 | 66 |
| H | 25 | 10 | 19 | 10 | 42 | 18 | 9 | 6 | 21 | 54 |
| LSD | 9 | 8 | 11 | 8 | 8 | 9 | 6 | 8 | 12 | 7 |

Notes: LSD is Fisher's least significant difference (P ≤ 5%). Sensory scale was from 0 (low/nil) to 100 (high).

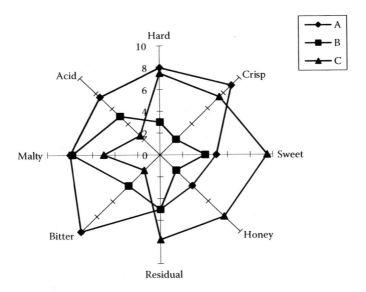

*Figure 2.10* Spider-web diagram summarizing the results of a descriptive analysis of three samples of biscuits.

Figure 2.10 is a spider-web diagram resulting from a descriptive analysis of three samples of biscuits. We prefer not to use this type of representation, as the lines are difficult to follow even for a reduced number of samples as shown in Figure 2.10. Figure 2.11 is a histogram representation of a descriptive analysis of three samples of mayonnaise. We find this type of graphical representation clearer than the spider-web graph, and it has the added advantage that LSD bars can be included to aid interpretation.

## 2.7.3   Affective tests

Affective tests are used to answer the question "Do consumers care about the difference?" (Table 2.3). This question is particularly relevant to SSL studies. As mentioned above, a food product close to the end of its SSL will not be exactly the same as the fresh product. A biscuit will have lost some of its crispness and powdered milk will have a slightly oxidized flavor. The question we want to ask the consumers is whether they care about these differences, that is, whether they still consider the product acceptable in spite of the sensory changes produced during storage. The following sections provide general guidelines for answering this question. Detailed examples of consumer tests conducted to estimate SSL will be given in subsequent chapters where the methodology is developed.

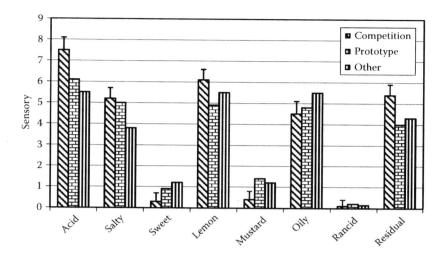

***Figure 2.11*** Sensory profile of three samples of mayonnaise. Bars represent Fisher's least significant difference.

### 2.7.3.1   Selecting consumers

When selecting consumers for an affective test, the central idea is to effectively sample the target population so that conclusions can be extended beyond the reduced number of consumers who actually taste the samples. There are books that provide comprehensive methodology for choosing an adequate sample from a population (Cochran 1977). When a company is about to launch a new product, managers understand that money has to be spent in conducting widespread, representative consumer tests. However, in the developing stage of a product, research and development (R&D) will probably have difficulty in convincing management that it is necessary to spend the money to conduct a representative consumer test in order to estimate the SSL of the product. In this last case, compromise solutions are sought, such as recruiting consumers among company employees or local residents. Even if this is the case, certain guidelines should be kept.

*Number of consumers.* Hough et al. (2006) performed calculations of the necessary number of consumers for consumer tests based on data from different countries and food products. Considering the average standard error, an alpha value (Type I error) of 5%, a beta value (Type II error) of 10%, and a difference between sample means of 10% of the sensory scale gave an N value of 112 consumers for this particular set of parameters. In the previous example N = 112 would be the necessary number of consumers in a study where each consumer measures all samples and the only

**Table 2.7** Classification of Frequency of Consumption
for Coffee and Spaghetti

| Classification | Coffee | Spaghetti |
| --- | --- | --- |
| Light | Less than 1 cup/day | Once/month |
| Moderate | 1–2 cups/day | 2–4 times/month |
| Heavy | More than 2 cups/day | 2 or more times/week |

interest is in determining significant differences between samples. If, for example, the acceptability of the samples was to be measured in two different locations within a country and the researchers wanted to compare the samples within each location, then N = 112 consumers would have to be used in each location, and that is a total of 224 consumers for the complete study. In Chapter 4 we will indicate the number of consumers necessary for SSL estimation based on survival analysis statistics.

*Frequency of consumption.* Generally consumers are classified as light, moderate, and heavy users. Table 2.7 classifies coffee and spaghetti according to frequency of consumption following general patterns in Argentina. The difference between these two products shows that what defines the actual frequency depends largely on the food in question. We would never expect to find a consumer who eats spaghetti as often as she drinks coffee! When market research staff is asked what frequency of consumption they aspire to in their consumer sample, their first answer will usually be, "We want heavy users!" Suppose the product is a certain brand of orange juice, and the heavy users are defined as drinking it every day. A simple task we perform in these cases is to take a survey of 30–50 employees or local residents near our institute and ask them how often they drink that brand of orange juice. The result of the survey may be 1 out of 30. Thus if 100 heavy users are to be recruited, approximately 3000 subjects will have to be interviewed. This is possible, of course, but costly. Very often the plan for a sample of heavy users is dropped to a sample of moderate users.

*Age.* A general age classification is children: 4 to 12 years; adolescents: 13 to 19 years; young adults: 20 to 35 years; adults: 36 to 65 years; and older adults: over 65 years. In our laboratory we are particularly wary about conducting consumer tests with children under the age of 10. To start with, children are usually given products they like: candy, chocolate, sausages, biscuits, ice cream, and so forth; thus, the chances are all samples will receive high scores. Another issue is that the test itself is exciting. Kids are taken to a special hall, samples are presented with mysterious three-digit codes in odd sorts of containers like small plastic cups or sample holders. They are told they are going to receive a present when they finish their task. All this setup is far removed from their everyday eating experience, and very likely the result will be far removed from what they really feel about the product in a real setting. Popper and Kroll (2005) reviewed the current state of knowledge in the area of consumer testing with children and stressed the importance of considering the sensory, cognitive, and social factors that may impact how best to conduct testing with this age group. In Chapter 5 (Section 5.3) a real case example will be presented on how SSL estimations can differ between children and adult populations.

*Gender.* In general, gender differences in food consumer habits tend to diminish. Decades ago certain products, such as whiskey, were targeted nearly exclusively at men, but these days men and women consume this type of product equally. Reduced calorie products are still mainly targeted at women, yet many men, due to overweight or other reasons, also consume these products. In our experience gender has never had an influence on SSL predictions.

*Income status.* This can be an important issue. We have observed that low-income children readily accepted aged yogurt samples that a population of middle-income adults widely rejected. In Argentina a practice of targeting certain income groups is to conduct a central location test in a district of a city where it is supposed that this group lives. We once conducted a study in a district of Buenos Aires where house owners were mostly upper-middle class. A hall was hired and potential consumers were interviewed as they walked down a pavement in the heart of this district. We soon realized that many of the women coming in to do the test were in fact the house servants of these upper-middle-class families who were out doing errands and could spend the time to do the test. In Argentina people are not willing to admit how much they earn. Due to this, some research agents ask subjects what car they have in their family, and this gives them a rough idea as to income status, probably better than other more intricate systems.

*Employees and local residents.* If the objective is to obtain a representative sample from the consumer population a product is targeted to, then the use of employees or local residents closely attached to the company is excluded. However, due to high costs or time constraints, companies often use employees or local population in certain stages of product development and/or maintenance. For the case of SSL estimations, it can be assumed that the majority of consumers will respond in a similar fashion. For example, in establishing the SSL of a biscuit, nearly all consumers will agree that they don't like a product that has lost its crispness; or most consumers don't appreciate seeing whey separation when they open a pot of yogurt. Thus it would seem reasonable to recruit consumers among employees or local residents for these cases. There could be a bias among company employees if they are very knowledgeable about the sensory properties of the fresh product, and thus any small changes will be detected and rejected, although these same small changes would not worry an average consumer. Another caveat in recruiting employees for SSL studies is that they could have knowledge of the objective of the experiment. If this were the case they could find all samples acceptable out of a sense of loyalty to their company. In conclusion, employees and local residents are a valuable source of subjects for consumer tests, but they should not be counted on before analyzing the possible biases they could introduce into the results.

### 2.7.3.2    Test location

Stone and Sidel (2004) presented a detailed discussion on the advantages and disadvantages of the three basic types of test locations: laboratory, central location, and home use.

*Laboratory.* In a laboratory study, employees or local residents are invited to sensory facilities in an institute or a company. Generally, these locations are of easy access, conditions such as sample presentation and ambience can be carefully controlled, and results can be analyzed and presented rapidly. Another advantage is that a relatively large number of samples can be served in a single session. For SSL studies, where we may wish a consumer to taste six or seven samples with different storage times, a consumer can taste three, take a 15-minute break where water and crackers are available as palate cleansers, and then taste the remaining three samples. This procedure is easily managed in a laboratory setup. The main disadvantage of the laboratory location is that standardized preparation procedures and product handling protocols might not necessarily mimic consumer behavior and experience at home. Another disadvantage is that the location suggests that the products are from the company and can thus generate biased results.

*Central location.* Tests at a central location are usually performed in a hired hall or the unused part of a cafeteria. Subjects are intercepted as they walk down a street pavement or the passageway of a shopping mall. They are asked screening questions, and if they meet the requirements for the test, they are invited to participate in a consumer test. Usually there is the enticement of a gift. Once in the hall they receive instructions and perform the test. The advantages of central location tests are that conditions can be controlled and that respondents are reasonably representative of the target population. One disadvantage is that conditions are artificial compared to real product usage at home, in a bar or restaurant, or at a party. The number of questions and/or products needs to be limited, as respondents are not willing to spend more than 15 or 20 minutes performing the test. For SSL studies, where it is customary to present six or seven samples to a consumer, 15–20 minutes would be sufficient if the questionnaire is kept very brief. Chapter 4 explains that for SSL estimation based on survival analysis statistics, the only answer we need from a consumer is whether she accepts or rejects the sample. If this were the case, a central location test would be appropriate.

*Home use tests.* Other tests are conducted at home in the last stage of the development of a food product to measure its performance under real-usage conditions. The other case where home use tests are called for is for products that are difficult to test elsewhere; for example, shampoo or a cake made from a packet where issues other than sensory are of importance. Advantages of home use tests are that the product is prepared and consumed under natural conditions of use, information can be gathered over repeated use of the product rather than first impression alone, and more information regarding the product in general can be obtained. For SSL studies the issue of consuming a small portion of an aged product in a laboratory or central location test (CLT) setting, versus the consumption of a regular portion in a home use test has not been researched. For example, for an SSL study on beer, will the probability of rejection of a 20-ml aged sample be the same as the probability of rejection of a whole bottle of the same sample? The disadvantages of home use tests are that they are time consuming, no more than two or three samples can be tested, and there is little control over preparation and consumption. This last issue can be of importance in SSL studies. For example, in an SSL study on cooked ham, consumers in the home use test should be asked to eat the ham as a cold meat accompanying a salad. Otherwise it might be consumed as a savory ingredient in a pie or sauce and thus, off-flavors of an aged sample could be unnoticed. Sanitary control is also of importance. In the cooked ham example, are refrigeration temperatures in consumer's homes to be trusted? Can it be ensured that they will consume the product within the time limits given to them in the instructions? The last thing we want is for consumers to get ill due to consumption of the samples they are asked to evaluate.

### 2.7.3.3  Quantitative affective test methods

Food preference that leads to food consumption depends upon characteristics of the individual, the food, and the environment (Shepherd and Sparks 1994). In SSL studies the focus is on the characteristics of the food, that is, appearance, texture, and flavor. However, other factors such as age or income status of the individual have been shown to affect the estimated shelf life of a product, as will be shown in Chapter 5.

Methods for measuring sensory acceptability were described by Meilgaard et al. (2007) and by Lawless and Heymann (1998). For SSL studies a very simple accept/reject question is used when applying survival analysis methodology (Chapter 4), and hedonic scale methods are used when applying the cut-off point methodology (Chapter 6). Regarding hedonic scales, the 9-point scale with phrases ranging from *dislike extremely* to *like extremely* is widely used and accepted among English-speaking populations. Curia et al. (2001) found that the scale's phrases did not translate well into Spanish, and they recommended the use of anchored box scales or numerical scales. Sosa et al. (2008) compared the use of box and number scales among low-income populations in Argentina, finding that both gave equivalent scores. However, after the sensory test, consumers were asked which scale they preferred, and 62% preferred the number scale, 33% the box scale, and 5% didn't prefer one more than the other. Also administrators of the tests received more questions as to how to complete the box scale in relation to the number scale. Thus the number scale was recommended. In Chapters 4, 5, and 6, examples of questionnaires with anchored box scales and number scales will be shown.

## References

Amerine, M.A., R.M. Pangborn, and E.B. Roessler. 1965. *Principles of sensory evaluation of food*, Chapter 9. New York: Academic Press.

ASTM E1871 Standard. 2006. *Standard guide for serving protocol for sensory evaluation of foods and beverages*. West Conshohocken: American Society for Testing Materials.

ASTM E1346 Standard. 2006. *Standard practice for bulk sampling, handling, and preparing edible vegetable oils for sensory evaluation*. West Conshohocken: American Society for Testing and Materials.

Barbieri, T., G. Hough, and H. Iglesias. 1999. Análisis sensorial de semillas de mostaza. In *Avances en análisis sensorial*, ed. M.H. Damasio and M.A. da Silva, pp. 139–147. San Pablo: Varela Editora.

Cochran, W.G. 1977. *Sampling techniques*. New York: John Wiley & Sons.

COI Norma T.20/Doc. no. 5. 2007. *Análisis sensorial del aceite de oliva: Copa para la degustación de aceites*. Madrid: Consejo Olivícola Internacional.

COI Norma T.20/Doc. no. 15. 2007. *Análisis sensorial del aceite de oliva: Método de valoración organoléptica del aceite de oliva virgen*. Madrid: Consejo Olivícola Internacional.

Curia, A.V., G. Hough, M.C. Martínez, and M.I. Margalef. 2001. How Argentine consumers understand the Spanish translation of the 9-point hedonic scale. *Food Quality and Preference* 12: 217–221.

Derndorfer, E., A. Baierl, E. Nimmervoll, and E. Sinkovits. 2005. A panel performance procedure implemented in R. *Journal of Sensory Studies* 20: 217–227.

Ellis, M.J. 1994. The methodology of shelf life estimation. In *Shelf life evaluation of foods*, ed. C.M.D. Man, and A.A. Jones, Chapter 2. Glasgow: Blackie Academic and Professional.

Garitta, L., G. Hough, and R. Sánchez. 2004. Sensory shelf life of dulce de leche. *Journal of Dairy Science* 87:1601–1607.

Hottenstein, A.W., R. Taylor, and B.T. Carr. 2008. Preference segments: A deeper understanding of consumer acceptance or a serving order effect? *Food Quality and Preference* 19: 711–718.

Hough, G., E. Martínez, and T. Barbieri. 1992. Sensory thresholds of flavor defects in reconstituted whole milk powder. *Journal of Dairy Science* 75: 2370–2374.

Hough, G., E. Martínez, T. Barbieri, A. Contarini, and M.J. Vega. 1994. Sensory profiling during ripening of Reggianito grating cheese, using both traditional ripening and in plastic wrapping. *Food Quality and Preference* 5: 271–280.

Hough, G., I. Wakeling, A. Mucci, E. Chambers IV, I. Méndez Gallardo, and L. Rangel Alves. 2006. Number of consumers necessary for sensory acceptability tests. *Food Quality and Preference* 17: 522–526.

IDF 99C Standard. 1997. *Sensory evaluation of dairy products by scoring*. Brussels: International Dairy Federation.

ISO 8586-1 Standard. 1993. *Sensory analysis: General guidance for the selection, training and monitoring of assessors, Part 1: Selected assessors*. Geneva: International Standard Organization.

ISO 11035 Standard. 1994. *Sensory analysis: Identification and selection of descriptors for establishing a sensory profile by a multidimensional approach*. Geneva: International Standard Organization.

ISO 13299 Standard. 2003. *Sensory analysis: Methodology: General guidance for establishing a sensory profile*. Geneva: International Standard Organization.

ISO 4120 Standard. 2004. *Sensory analysis: Methodology: Triangle test*. Geneva: International Standard Organization.

ISO 5495 Standard. 2005. *Sensory analysis: Methodology: Paired comparison test*. Geneva: International Standard Organization.

ISO 8589 Standard. 2007. *Sensory analysis: General guidance for the design of test rooms*. Geneva: International Standard Organization.

ISO 11132 Standard (working draft). 2008. *Sensory analysis: Methodology: General guidance for monitoring the performance of a quantitative sensory panel*. Geneva: International Standard Organization.

Kim, H., K. Kim, S.Y. Jeon, J. Kim, and M. O'Mahony. 2006. Thurstonian models and variance II: Experimental confirmation of the effects of variance on Thurstonian models of scaling. *Journal of Sensory Studies* 21: 485–504.

Lawless, H.T., and H. Heymann. 1998. *Sensory evaluation of food, principles and practices*. New York: Chapman & Hall.

Lea, P., T. Næs, and M. Rodbotten. 1997. *Analysis of variance for sensory data*. Chichester, UK: John Wiley & Sons.

MacFie, H.J., N. Bratchell, K. Greenhoff, and L.V. Vallis. 1989. Designs to balance the effect of order of presentation and first-order carry-over effects in hall tests. *Journal of Sensory Studies* 4: 129–148.

MacFie, H. 2007. Preference mapping and food product development. In *Consumer-led food product development*, ed. H.J.H. MacFie (p. 562). Cambridge, UK: Woodhead Publishing Limited.

Martínez, C., M.J. Santa Cruz, G. Hough, and M.J. Veja. 2002. Preference mapping of cracker type biscuits. *Food Quality and Preference* 13: 535–544.

Mead, R. 1988. *The design of experiments: Statistical principles for practical applications*, pp. 407–410. Cambridge, UK: Cambridge University Press.

Meilgaard, M.C., Civille, G.V., and B.T. Carr. 2007. *Sensory evaluation techniques*. Boca Raton: CRC Press.

Meyners, M., and P.B. Brockhoff. 2003. The design of replicated difference tests. *Journal of Sensory Studies* 18: 291–324.

Moskowitz, H.R., and B. Krieger. 1995. The contribution of sensory liking to overall liking: An analysis of six food categories. *Food Quality and Preference* 6: 83–90.

O'Mahony, M. 1986. *Sensory evaluation of food: Statistical methods and procedures*. New York: Marcel Dekker, Inc.

O'Mahony, M., and B. Rousseau. 2002. Discrimination testing: A few ideas, old and new. *Food Quality and Preference* 14: 157–164.

Popper, R., and J. Kroll. 2005. Conducting sensory research with children. *Journal of Sensory Studies* 20: 75–87.

Poste, L.M., D.A. Mackie, G. Butler, and E. Larmond. 1991. *Publication 1864/E1991: Laboratory methods for sensory analysis of food.* Ottawa: Research Branch Agriculture Canada.

Presa-Owens, C., and A.C. Noble. 1997. Effect of storage at elevated temperatures on aroma of Chardonnay wines. *Am. J. Enol. Vitic.* 48: 310–316.

Ramírez, G., G. Hough, and A. Contarini. 2001. Influence of temperature and light exposure on sensory shelf life of a commercial sunflower oil. *Journal of Food Quality* 24: 195–204.

Shepherd, R., and P. Sparks. 1994. Modelling food choice. In *Measurement of food preferences*, ed. H.J.H. MacFie and D.M.H. Thomson, Chapter 8. Glasgow: Blackie Academic and Professional.

Sosa, M., C. Martínez, F. Márquez, and G. Hough. 2008. Location and scale influence on sensory acceptability measurements among low income consumers. *Journal of Sensory Studies* 23: 707–719.

Stone, H., and J.L. Sidel. 2004. *Sensory evaluation practices, 3rd edition.* San Diego: Elsevier Academic Press.

Van Toller, S.V. 1993. The psychology and neuropsychology of flavor. In *Flavor measurement*, ed. C.T. Ho and C.H. Manley, Chapter 10. New York: Marcel Dekker.

# chapter 3

# Design of sensory shelf-life experiments

## 3.1 Initial considerations

Following are characteristics of food products and their modes of deterioration along with corresponding examples:

1. *Heterogeneous:* yogurt with fruit pulp, ravioli
2. *The chemical reactions involved in their deterioration have complex kinetics:* Maillard reaction, lipid oxidation
3. *Rheological and textural properties are not simply mod*eled: non-Newtonian fluids, viscoelasticity
4. *Biological changes continue during food storage*: enzymes that keep active even under frozen conditions; products with live microorganisms, like cheese
5. *The deterioration of a food during storage is often the result of a number of simultaneous processes:* growth of microorganisms, enzymatic and non-enzymatic reactions, physical processes such as moisture loss

These characteristics make food, generally speaking, a difficult product to study. Added to this are expectations of today's consumers. On the one hand, they want sensory quality, convenience, nutrition, year-round availability, and prolonged storage; on the other hand, they also want fewer additives, organic-type practices, and minimum processing. A difficult proposal indeed!

The shelf life of a food product will depend on the following:

- *Formulation:* high-quality ingredients, moisture content, pH, and preservatives
- *Process:* must limit deterioration yet be favorable to desired properties
- *Packaging:* must create the appropriate storage atmosphere within the package—an adequate balance of oxygen, carbon dioxide, and/or inert gases—and should also provide resistance to mechanical stress
- *Storage conditions:* must carefully control humidity, lighting, and temperature—the factors that most frequently accelerate food deterioration

## 3.2 Approximations of shelf-life values

Following are some alternatives for measuring approximate shelf-life values. However, these alternatives are no substitute for a properly designed and conducted shelf-life study.

### 3.2.1 Literature values

In Chapter 1 a review of the most relevant books dealing with shelf life of food products was presented. In some of these books, shelf-life values at the time of publishing are documented. For example, Figure 3.1 shows days of shelf life based on unacceptable flavor for nonfat dry milk. Suppose it is of interest to estimate shelf life at 20°C for dry milk with 5% moisture content; Figure 3.1 shows this is approximately 450 days. The following caveats should be considered regarding this value:

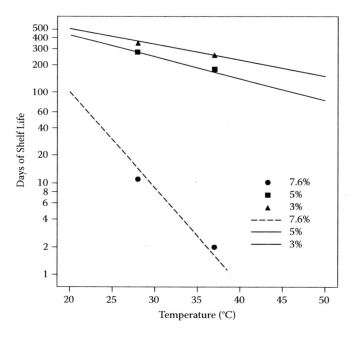

**Figure 3.1** Days of shelf life for nonfat dry milk stored at different temperatures (°C) and different moisture contents expressed as %. (Reprinted with permission from: Labuza, T.P. 1982. *Shelf-life dating of foods*. Westport: Food & Nutrition Press.)

- The shelf-life value at 20°C is extrapolated from measurements taken at 28°C and 37°C. In Chapter 7 the high variability inherent in extrapolations based on only two temperature points will be discussed. The true shelf-life value could easily be half or double the value read from Figure 3.1.
- There is no indication as to how the "unacceptable flavor" was defined or measured.
- The drying procedure and packaging used during storage are not specified.

Martin (1994) presented a table with "representative shelf life of some chocolate confectionary." For example, the shelf life of milk chocolate is given as 16 and 12 months for temperate and tropical conditions, respectively. According to the author, these would be representative values for products adequately wrapped to withstand the effects of high relative humidity and to exclude light. These values can only be taken as general guidelines as there is no indication of formulation, processing, or exactly what wrapping was considered. Also, storage temperature conditions are vaguely defined.

The two examples presented are representative of the literature on shelf values for food products. These approximate values can be useful to design a shelf-life study of a product of interest, for example, to define maximum storage times for the study and corresponding temperature conditions.

## 3.2.2   Values from the Internet

The Internet is a source of information that could be consulted for shelf-life values. Following are two of many sources to be found:

- http://www.amarogue.com/foodcode.html
- http://whatscookingamerica.net/Information/FreezerChart.htm

Web pages soon become outdated, so there is no guarantee that the above sites will be available when this book is published. However, the type of information contained in Web pages like these is shown in Table 3.1. A wide shelf-life range is given for some products like butter: 1 to 3 months. In Argentina leading processors give pasteurized milk and hot dogs shelf lives of 14 days and 30 days, respectively, far longer than the values shown in Table 3.1 Also, details important to shelf life, such as the gas under which the olive oil was bottled, are not indicated. Therefore, the Internet would not seem an adequate source for food shelf-life values.

*Table 3.1* Shelf Lives of Certain Food Products

| Product | Storage temperature | Shelf life | Comments |
|---------|---------------------|------------|----------|
| Cheesecake | 3–4°C | 3–7 days | Wrap well |
| Butter | 3–4°C | 1–3 months | Leave in original wrapping |
| Milk | 3–4°C | 7 days | |
| Hot dogs | 3–4°C | 2 weeks | Lose flavor quickly |
| Flour | Room | 6–8 months | |
| Olive oil | Room | 24 months | |

*Note:* http://whatscookingamerica.net/Information/FreezerChart.htm (accessed April 2009).

## 3.2.3   Values based on distribution times

Distribution times may be considered by some as a means of estimating shelf life. For example, a small- or medium-sized company could decide to produce a chocolate biscuit similar to a leading brand in formulation and packaging to be marketed at a lower price. Instead of conducting a shelf-life study, they could decide to use the same *best before* period as the leading brand. Once the product is on the market, they would check on complaints, hoping none appear. A number of reasons could cause the product's shelf life to differ from that of the leading brand. The leading brand could have

- An antioxidant that prevents the onset of oxidized flavors
- Improved packaging technology that guarantees dry air within the package and proper closures
- Positive expectations among consumers who tend to assimilate small changes occurring during storage in the leading brand but not in a lesser known cheaper brand

We were once contacted by a mayonnaise producer that assigned its product 6 months' shelf life and discovered that its main competitors had extended shelf life to 8 months. Marketers convinced management that they could not have a shorter shelf life, and thus the *best before* period was changed to 8 months. This was when they started receiving consumer complaints of products after 5.5 to 8 months storage. We were asked to conduct a sensory shelf life (SSL) study, and the result was an estimated shelf life close to 6 months. Marketing and management realized their mistake and brought the *best before* period back to 6 months until a change (formulation, ingredient quality, manufacturing practice or packaging) followed by an SSL study justified an extension of the *best before* period. It should be noted that when the *best before* period was 8 months, there were complaints of products with 5.5 months storage; yet when the period was 6 months, these complaints were rare. Two reasons for this are (1) manufacturers remove products from the shelf before expiration date; thus, consumers would not usually find a 5.5-month-old product when the *best before* period is 6 months; and (2) consumers are reluctant to buy products close to their expiration date; if the *best before* period is 8 months, a product with 5.5 months is likely to be on the shelf and is likely to be bought.

A rather haphazard method to determine the shelf life of a food product would be to roughly estimate a value based on previous knowledge and similar products in the marketplace, use this estimated value, and register consumer complaints. If no complaints associated to shelf life come in, it can be assumed that the estimation was valid or could even have been an underestimation. If complaints do come in, then *best before* period is reduced. The main drawback of this trial and error methodology is that errors in shelf-life estimation are detected by the consumer and thus serious damage can be done to the brand and manufacturer's image. Another drawback is that when a complaint is made, rarely are storage conditions or modes of deterioration known. A consumer can report a stale flavor in a cake mix powder, informing that it had been stored in a cupboard in the kitchen, but omitting to mention that the cupboard was next to the kitchen stove with no insulation between the two components. Thus the cake mix powder was subjected to temperatures between 35°C and 40°C for several hours each day.

## 3.3  Temperatures and storage times

### 3.3.1  Temperatures

Taoukis et al. (1997) provide some guidelines for possible storage temperatures for different food categories. For example, for dehydrated foods they recommend storage temperatures of 25, 30, 35, 40, and 45°C; and storage of the control sample at –18°C. For frozen foods they recommend –5, –10, and –15°C; and control at –40°C. These suggested temperatures should be analyzed in each case. For example, for powdered milk, which is a dehydrated product, storage above 35°C is not recommended because dairy fat starts melting at approximately 37°C, leading to a different deterioration mechanism from what occurs at room temperature. A note on *room temperature*: This term should always be accompanied by the degrees Celsius it represents; for example: "Samples were stored at a room temperature of 20°C ± 1°C."

## 3.3.2   Maximum storage time

When designing an SSL study, one of the first parameters to be estimated is the maximum storage time the study will cover. For this the following should be considered:

- A first approximation can be obtained from the company's experience with the product or similar products. Taking note of the *use before* periods used by competitive brands can also be of help. Another possibility is to search stores for products of competitive brands close to their expiration date; these can be compared with fresh products of the same brand, and if deterioration is present, the period can help in estimating maximum storage time.
- Some companies have a storage bank, which is a room or controlled chamber where daily or weekly samples are kept for a certain time. These storage banks are of great value in obtaining initial estimates for shelf-life studies. On some occasions, conditions in these storage banks are well controlled, and the samples stored there are perfectly appropriate to be used in an SSL study. A problem we have encountered is that rarely is the sample quantity stored in these banks sufficient for the consumer tests necessary to obtain SSL estimates.
- If previous knowledge of the product's shelf life is limited, and thus a maximum storage time cannot be reasonably estimated, the only solution is to conduct a preliminary study. This would consist of storing the product at a relatively high temperature and then extrapolating the results to estimate a maximum storage time at room temperature. For example, if a company produces a new salad dressing and has no previous knowledge of its shelf life, they can store samples at 45°C and control samples at 4°C. Every 3–4 days they can test a sample stored at 45°C versus the control, using a difference from control scale (see Section 2.7.1.4). For these preliminary studies a reduced panel of three or four assessors can be used to arrive at a result by consensus. Suppose that at 45°C the salad dressing develops a clear off-flavor after 25 days' storage. A general value of $Q10 = 2.5$ can be adopted (see Section 7.3); and thus at a storage temperature of 25°C, an approximate maximum storage time would be 156 days. It would be wise to store enough samples at 25°C to cover 7 months' storage as the 156-day estimate is prone to error.
- For SSL models to have good fits and thus good predictive power, the products in an SSL study should reach considerable sensory deterioration. When the maximum storage time has been reached, the product should have a high percentage of rejection associated with clear development of sensory defects.

- Associated with the above recommendation that a high percentage of rejection be reached is an ethical consideration. If a trained assessor or a consumer is to taste 20 ml of a highly oxidized reconstituted milk powder, there is no health hazard involved, but no doubt there will be complaints about the ghastly flavor of the sample. Is it ethical to ask a person to taste a sample that is extremely unpleasant? It could be argued that this is necessary once to establish a good correlation between a sensory and a chemical measurement so that in the future this last index can be used, thus avoiding future unpleasant sensory experiences. As mentioned in Chapter 2, in our department we have an ethics committee formed of the project leader, a trained sensory assessor, and a researcher from another department of our institute. The project is presented together with representative samples of what is likely to come up during the development of the project. The committee may decide to consult toxicological information regarding the samples, and the slightest doubt in this respect leads to a project cancellation or reformulation. The committee can also decide on a reformulation if they consider that the samples are beyond what is reasonable to ask a person to taste. The project leaders know they have to present their projects to the committee, so they take care of ethical issues, and thus it is very rare for a project to be turned down.

## 3.3.3   Time intervals

Figure 3.2 shows a hypothetical experiment. If only three time intervals were taken and the results were those represented by the square symbols,

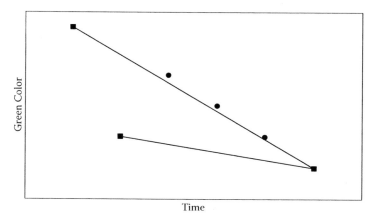

***Figure 3.2*** Green color of a vegetable versus storage time.

the tendency of the green color over time could not be established. Only if the experiment were completed with three other time intervals represented by the round symbols would it be clear which of the original points was an outlier. Even if three points are reasonably well aligned, the predictive ability of a regression with only three points is very low. A minimum of six time intervals is recommended.

Suppose an SSL study on ultrapasteurized milk with a maximum storage time of 21 days at 5°C and eight time intervals. One possibility would be to distribute the time intervals equally, that is, to perform sensory measurements at days 0, 3, 6, 9, 12, 15, 18, and 21 days. It is probable that up to 9 days there would be very small if any sensory changes; thus, measurements at days 3 and 6 would be irrelevant. In this case a more reasonable time distribution would be 0, 9, 12, 15, 17, 18, 19, and 21 days. Knowledge of the kinetics of the sensory changes that occur during storage is valuable in deciding on the time intervals. In Chapter 6 (Section 6.5) zero-order and first-order kinetics will be discussed, as well as a lag phase in the development of a sensory defect. Gacula (1975) presented designs where certain rules are followed in establishing the time intervals. In our practice we have found that knowledge of kinetics and common sense are more helpful in establishing time intervals than strict rules. Common sense means accounting for weekends or odd times of the day when participation of trained assessors or consumers is complicated.

### 3.3.4   Critical descriptor

When a food product undergoes prolonged storage, sensory changes can be defined by one or more descriptors. The critical descriptor is the one that limits the SSL. Undesirable sensory descriptors increase during storage. Examples are the browning of sweetened condensed milk or the oxidized flavor of lipid-containing products such as mayonnaise. Desirable sensory descriptors can decrease during storage, and examples are the loss of crispness in biscuits or the loss of green color in vegetables such as broccoli.

A critical descriptor can be the one that changes at the highest rate during storage. Martínez et al. (1998) studied the shelf life of mayonnaise and found that total aroma, egg, oily and oxidized flavors changed during storage; and they determined that oxidized flavor was the critical descriptor because it was the one that changed at the highest rate.

Another criterion to define a critical descriptor is to consider the one most important to the consumer. Curia and Hough (2009) studied the SSL of a human milk replacement formula designed for babies in their early months of age. A survey among mothers who were giving this type of formula to their babies showed that mothers rarely tasted the product, and thus the salient sensory attribute they would pay attention to was color when preparing their baby's bottle. If a portion was rejected by the babies, mothers attributed the rejection to stomach trouble and not to inappropriate flavor. Thus the color was the salient attribute that mothers would pay attention to and was thus considered the critical descriptor. An accelerated storage test also showed that color increased at a higher rate than oxidized flavor. Another situation is when a sensory defect does not worry consumers. Hough et al. (2002) found that an increase in cooked flavor in reconstituted whole milk powder did not affect acceptability, not even for a sample placed in a boiling water bath for 15 minutes. For caramel flavor, acceptability only decreased for a sample with the highest concentration of flavoring, well beyond what would be expected under normal circumstances. Their conclusion was that from a consumer's point of view, these descriptors would not normally be critical in defining the SSL of milk powder.

A critical descriptor can also be defined by the concept of a product. Restrepo et al. (2003) showed that coffee consumers, both in Argentina and Colombia, tolerate a certain level of oxidized flavor without diminishing their acceptability scores. However, the institution that controls the quality of Colombian coffee, the Federación Nacional de Cafeteros, regardless of whether consumers mind, requires that premium coffee should have an absence of oxidized notes as measured by their trained sensory panel (personal communication, 2002). This criterion defines the critical descriptor for this product from a conceptual point of view.

### 3.3.5 Storing fresh samples

When conducting an SSL study it is always of interest to compare samples of different storage times with a sample that is considered fresh. For reversed storage designs (see Section 3.3.6.2), being able to maintain samples without alterations is central to the design.

For some SSL studies keeping a fresh sample is relatively simple. Martínez et al. (1998), in their study on SSL of mayonnaise, compared the stored samples with their corresponding controls. The controls belonged to the same lots as the samples under shelf-life study but were kept in refrigerated storage at 5°C. It was considered that sensory changes at this temperature were insignificant compared with changes at shelf-life study temperatures (>20°C).

In other cases keeping a fresh sample is almost impossible. Araneda et al. (2008) applied survival analysis statistics to estimate the SSL of ready-to-eat lettuce stored at 4°C, and for a 50% rejection probability shelf life ± 95% confidence limits was 15.5 ± 0.9 days. They stored a single batch and at predetermined storage times consumers evaluated the lettuce samples. Due to their design of the study, known in survival analysis statistics as *current status data*, no control sample was necessary. However, if they would have wanted consumers or a trained panel to compare stored samples with fresh, where would they have stored the fresh sample? Freezing a batch of lettuce would not have helped, as this process alters the texture considerably. Storing a batch of lettuce just above 0°C to prevent freezing would have retarded deterioration, but this batch would not be completely unaltered in relation to the batch stored at the study temperature of 4°C. One possibility would be to harvest a fresh batch at each storage time, but lettuce, like many other vegetable products, varies from one batch to another. When evaluating lettuce samples with different storage times, each time comparing with a different fresh batch, the differences between storage times would be confused with the differences between batches. These possibilities are summarized in Table 3.2. Thus, for a product like lettuce, it is difficult if not impossible to have an adequate fresh sample for comparison purposes. When a consumer goes shopping for lettuce, he has an internal standard formed by previous experience of what he considers to be fresh lettuce; thus, the lack of a physical sample of fresh product would not be a problem as it does not resemble a usual shopping experience. In the case of a trained panel, the lack of a physical standard means they have to rely on the training they received on the sensory characteristics of a fresh sample.

In other SSL studies, keeping a fresh standard has intermediate difficulty. Curia et al. (2005) studied the SSL of commercial yogurt. Bottles (1000 ml) from different batches were stored at 10°C in such a way as to have samples with different storage times ready on the same day. Storage times at 10°C were 0, 14, 28, 42, 56, 70, and 84 days. All batches were made with the same formulation and were checked to be similar to the previous batch by consensus among three expert assessors. To perform this checking, the previous batch was stored at 2°C; changes at this temperature over 14 days were considered negligible in relation to changes at 10°C. One difficulty with this alternative of using different batches over time is if there is a drift in the sensory properties of the product. This would mean that changes between consecutive batches are small, yet they are

*Table 3.2* Alternatives of Keeping or Obtaining a Fresh Batch of Lettuce to Compare with a Batch Stored Different Times at 4°C

| Alternative | Consequence |
| --- | --- |
| Freeze at –18°C | Considerable texture change |
| Store at 0°C | Changes are slower than at 4°C but cannot be disregarded |
| Harvest a fresh batch at each storage time | Fresh batches vary from one to another, and thus differences due to storage time would be confused with differences between batches |

appreciable between the initial batch and the final batch used for a shelf-life study. If this happens, it is due to poor quality control, and this department's methodology should be revised.

## 3.3.6    Basic and reversed storage designs

In this section alternative strategies for storing and retrieving products will be presented.

### 3.3.6.1    Basic design

This design is the first one that comes to mind when thinking about conducting an SSL test. It consists of storing a single batch at the desired temperature and periodically removing samples from storage and analyzing them. Figure 3.3 shows this design for an SSL study on mayonnaise stored at 20°C. There are two major drawbacks to this basic design.

One drawback is that for each storage time a corresponding sensory analysis has to be performed. For the example shown in Figure 3.3, and depending on the type of study, this would mean assembling a trained panel, a consumer panel, or both on six separate occasions. For a trained panel, between each occasion and the next, as in the example of Figure 3.3, the time elapsed means introducing one or two warm-up sessions to make sure the panel remembers descriptors and their references. With a trained panel, ideally all assessors would remain on the panel throughout the study. If an assessor was absent for one of the storage assessments, his missing values can be dealt with. Statistical programs such as Genstat (VSN International, Hemel Hempstead, UK) estimate the most probable missing values when performing an analysis of variance; if the design becomes too unbalanced, alternative forms of analysis such as residual maximum likelihood (REML) technique can be used (Horgan and Hunter 1993). With consumer panels, assembling the same consumers for each of the storage times can be complicated and costly. When analyzing consumer data it is generally assumed that each consumer conforms a block in the design; that is, the perception and outlook of a certain consumer are the same for each sample she evaluates. For a study such as the one shown in Figure 3.3, more than 7 months elapse between the first and last evaluation. A consumer could have been recruited as a regular mayonnaise consumer, but during the 7-month period may have stopped consuming the product. This could lead to a change in consumer acceptance due to changing habits and not due to storage times. Also, consumers get to know they are on a mayonnaise study and very probably start paying attention to their everyday consumption of the product, thus gaining a certain degree of expertise, which is undesirable in a consumer study. To avoid these problems with consumer panels when using the basic design,

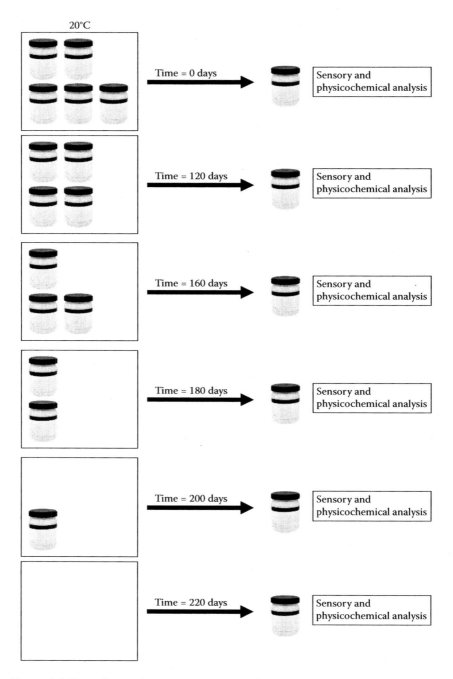

*Figure 3.3* Basic design for mayonnaise stored at 20°C.

in Section 5.2 a method will be proposed by which each consumer evaluates samples corresponding to a single storage time.

Another drawback to the basic design is that trained assessors and/or consumers can become aware that they are participating in an SSL study. This can lead to biased results. If a trained assessor finds today's mayonnaise sample with an oxidized score of 5 on a 1–10 intensity scale, then 20 days later an expectation error can lead him to score the sample with intensity above 5. Analogously, if a consumer rejects today's sample, when he is asked to come in again to taste mayonnaise in 20 days, expectation will lead him to reject the sample also. There are ways around these expectation errors, for example, an extra session in which samples taken from a fresh batch of mayonnaise are presented to the trained panel and/or consumers. Once everyone has finished their evaluations, their results can be discussed with everyone knowing what samples they had evaluated.

An advantage of the basic design is that it is not necessary to store a fresh sample. In Section 3.3.5 cases were presented where the storage of a fresh sample was particularly complicated. If a fresh sample can be stored without too much difficulty, it is advisable to do so; especially if a trained panel is going to be evaluating stored samples. In this case it is easier for the trained assessor to compare the stored samples versus the fresh sample. Ramírez et al. (2001) studied the SSL of sunflower oil at 35°C, 45°C, and 60°C, both in the dark and with 12 hours of daily illumination, using a basic design. A fresh sample was kept at 4°C in the dark. For each storage time the trained panel compared the stored sample with the fresh sample.

### 3.3.6.2   Reversed storage design

The basic idea of the reversed storage design is to have all samples, each with a different storage time, all available on the same day. Figure 3.4 shows this design for an SSL study on mayonnaise stored at 20°C. It is assumed that changes at 4°C are negligible. As can be seen, the sample that went into 20°C storage at 0 days will have the longest storage time at 20°C: 220 days. The sample that went into 20°C storage at 20 days will have 200 days' storage at the end. The sample that remained at 4°C during the whole period will be considered as the fresh sample because it was never at 20°C.

The big advantage of the reversed storage design is that all samples are available for evaluation at the same time. If these samples are to be evaluated by consumers, it means recruiting the consumers on a single occasion to evaluate all samples. In the mayonnaise example this would mean evaluating six samples, which can easily be handled by a consumer with sufficient time between samples.

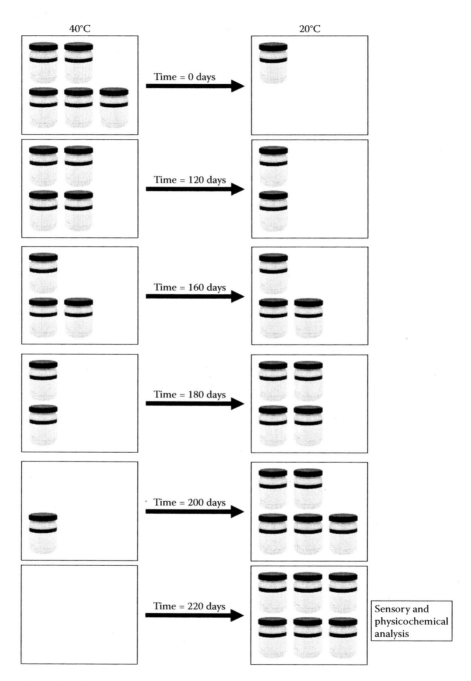

*Figure 3.4* Reversed storage design for mayonnaise.

For a trained panel, it is far more efficient to train the participants once and evaluate the samples than to repeatedly train the panel to evaluate each sample separately as in the basic design. In Section 2.4.7 the number of samples to be evaluated in a session was discussed. For the mayonnaise example, the number of samples per session would depend on the number of descriptors. If there are three or four mouth texture descriptors and four or five flavor descriptors, an assessor would have to taste each sample three or four times. If this was the case, no more than four samples per session would be recommended. Having six samples to evaluate, if the evaluations are done by duplicate, would mean a total of three sessions. For the mayonnaise example, when the storage is over, that is, when day 220 has been reached (last box in Figure 3.4), all samples should be stored at 4°C till evaluation. This storage period at 4°C should be as short as possible: 3–4 days at the most. At 20°C oxidation reactions would have started and cooling down to 4°C will slow these reactions down but by no means freeze them.

If fresh bread is frozen and then defrosted, it regains its freshness. However, its shelf life is shorter than bread that has not undergone the frozen–defrosted cycle. Giménez et al. (2007) applied different approaches to estimate the sensory shelf life of brown pan bread treated with different enzymes. Breads were stored in a temperature-controlled storage room at 20°C for 1, 4, 7, 10, 13, 15, and 17 days. After reaching the desired storage times, breads were frozen at –20°C and stored at –18°C, providing samples with different storage times from one batch. The samples were defrosted at 20°C for 6 hours previous to their evaluation. This storage scheme is shown in Figure 3.5. There must be reasonable certainty that the frozen–defrosted cycle does not change sensory properties per se. In our laboratory we had to compare two samples of cracker-type biscuits. Sample A had to be compared with Sample B, which was going to be available 3 months later. Could Sample A be frozen during those 3 months without its sensory properties changing? We took three packets of Sample A and froze them for 48 hours and then defrosted them for 6 hours at 20°C. These crackers that had undergone the frozen–defrosted cycle were compared with other crackers of Sample A that had not undergone the cycle using a triangle test. No significant differences were found, and thus we were reasonably certain that the crackers could be frozen without perceptible sensory changes.

The reversed storage design is not always possible or convenient. This occurs when there is difficulty in storing samples in conditions where sensory changes do not occur or occur very slowly. For the mayonnaise example presented in Figure 3.4 unchanging samples were stored at 4°C; and for the bread sample presented in Figure 3.5 unchanging samples were stored at –18°C. In Section 3.3.5 the storage of a control sample was discussed. Lettuce was presented as an example in which this storage is

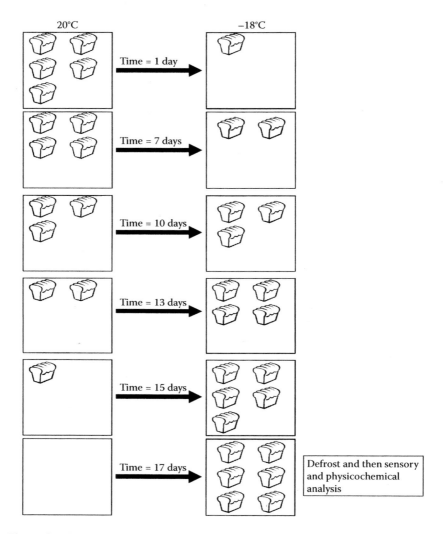

*Figure 3.5* Reversed storage design for pan bread.

particularly difficult. If a storage condition where the sample remains unchanged cannot be obtained, an alternative for a reversed storage design would be to introduce samples from different production batches over the designed storage time. This can be done if there is certainty that the variations between batches are sufficiently small. Otherwise the batch variations will be confused with the storage time variations. In the cases in which the reversed storage design is not possible, the basic design (Section 3.3.6.1) or current status data design (Section 5.2) has to be used.

## 3.3.7  How much sample should be stored for sensory shelf life studies?

When performing an SSL study a practical issue is calculating the sample quantity that has to be stored to cover the whole study. To illustrate calculations three case studies will be presented.

### 3.3.7.1  Bananas

Very often consumers buy bananas that are too green to be consumed immediately; they take these bananas home, store them at room temperature for a few days, and then start consuming them. If the bananas are kept too long, they will get overripe. Thus, for this product there are two events of interest, which are shown in Figure 3.6. The model to be used to determine optimum storage time is presented in Chapter 8 (Section 8.5). Bananas are stored at 20°C, and for each storage time consumers are asked to respond a three-category question:

- Is the banana too green?
- Is the banana okay?
- Is the banana overripe?

Suppose the following for the study:

- A total of eight storage times cover the span from too green to overripe.
- For each storage time 80 consumers evaluate the bananas.
- Each consumer receives a single banana for appearance. This banana is removed, peeled in a preparation room, and presented to the consumer with a different code for texture and flavor evaluation.

With these considerations a minimum of 8 storage times × 80 consumers = 640 bananas are necessary. At each storage time the 80 bananas to be presented to consumers have to be chosen to be approximately uniform in their ripening stage. This can be done by the panel leader with the help of two assistants. Experience tells the panel leader that to have 80 uniform bananas, no fewer than 120 have to be available. Thus, a total of 8 storage times × 120 = 960 bananas necessary. To be on the safe side, 1200 bananas are stored. If this quantity of bananas is out of the question, for example, because there is not enough room for

*Figure 3.6* Two events of interest for ripening of bananas.

them in the 20°C storage chamber, a solution could be to have groups of four consumers at a table and all evaluate a single banana for appearance. This banana would then be removed, peeled, and cut in four portions for texture and flavor evaluations; these portions would be served to the consumers with different codes. There is a risk of interaction among the consumers during the evaluation. Even if they are asked to perform the task individually the temptation of peeping at another person's score is always present.

### 3.3.7.2   Vegetable oil

An SSL study is to be performed on vegetable oil bottled in plastic PET bottles in an atmosphere of nitrogen. The cutoff point methodology is to be applied; this method will be covered in detail in Chapter 6. Samples during storage are evaluated by a trained sensory panel of 10 assessors. Suppose the following for the study:

- The temperatures of interest are 35°C, 45°C, and 60°C.
- Two illumination conditions are considered: darkness and fluorescent daylight-type illumination 12 hours a day.
- The smallest-sized bottle that can be produced at the company has a 500 ml capacity.
- The fresh control is kept at 4°C in darkness.
- For each storage temperature and illumination condition, there will be six storage times
- A reversed storage design will be used following the scheme presented in Figure 3.4.
- For each storage temperature and illumination condition, there will be six samples for evaluation, each one corresponding to a different storage time.
- Each sample will be 20 ml and served in a covered 70-ml glass.
- Measurements will be performed in duplicate. Although a single bottle (500 ml) would be enough for 10 assessors to measure each sample in duplicate (10 assessors × 20 ml × 2 = 400 ml), it is decided to use a different bottle for each repetition.
- The trained sensory panel will compare each sample's oxidized flavor with a fresh control using a structured scale of 1–10 points anchored with *equal to control* at one extreme and *a lot more than control* at the other extreme. At a single session they will evaluate the six samples corresponding to a storage temperature and illumination condition.
- Apart from the sensory measurements, peroxide and anisidine values are to be measured, for which a minimum of 200 ml is necessary.

From above considerations, the number of bottles to be stored at each condition is as follows:

- Sensory analysis: 6 storage times × 2 repetitions = 12 bottles
- Physicochemical analysis: 6 storage times = 6 bottles
- Total number of bottles for each condition = 18 bottles
- To be on the safe side, 24 bottles are required for each condition.

Therefore, the total number of bottles is: 3 temperatures × 2 illuminations × 24 bottles = 144 bottles. The total number of fresh control bottles necessary for the sensory analysis is 3 temperatures × 2 illuminations × 2 repetitions = 12 bottles. To be on the safe side, 20 bottles are stored at 4°C in the dark.

### 3.3.7.3    Yogurt
The SSL of yogurt stored at 10°C is to be studied. Suppose the following for the study:

- There will be six storage times between 7 and 62 days.
- For each storage time the company prepares a new batch of yogurt. It is supposed that differences between batches are small. Thus it is a type of reversed storage design, similar to the scheme shown in Figure 3.4.
- At the end of the storage time, that is, after 62 days, samples are removed from the 10°C refrigerator and stored at 2°C for a maximum of 2 days till measurement.
- Samples will be measured by 100 consumers and in duplicate by a 12-member trained panel.
- Each consumer and each trained assessor will receive six pots of yogurt, each with their corresponding storage time, in random order. The pots all have the same brand and are coded with 3-digit numbers. The lids, which have the *best before* dates, are removed. Each pot contains 125 g of yogurt.
- Both for consumers and for trained assessors the whole-pot presentation is considered necessary for an adequate evaluation of free whey produced during prolonged storage.

From these considerations the number of pots to be produced for each storage time is as follows:

- 100 for consumers + 24 for trained assessors (12 assessors × 2 repetitions) = 124 pots
- To be on the safe side, 160 pots are manufactured for each storage time.

Considering the six storage times, there will be total of 6 times × 160 pots = 960 pots.

# References

Araneda, M., G. Hough, and E. Wittig de Penna. 2008. Current-status survival analysis methodology applied to estimating sensory shelf life of ready-to-eat lettuce (lactuta sativa). *Journal of Sensory Studies* 23: 162–170.

Curia, A., M. Aguerrido, K. Langohr, and G. Hough. 2005. Survival analysis applied to sensory shelf life of yogurts. I: Argentine formulations. *Journal of Food Science* 70: S442–S445.

Curia, A., and G. Hough. 2009. Selection of a sensory marker to predict the sensory shelf life of a fluid human milk replacement formula. *Journal of Food Quality* 32: 793–809.

Gacula, M.C. 1975. The design of experiments for shelf life study. *Journal of Food Science* 40: 399–403.

Giménez, A., P. Varela, A. Salvador, G. Ares, S. Fiszman, and L. Garitta. 2007. Shelf life estimation of brown pan bread: A consumer approach. *Food Quality and Preference* 18: 196–204.

Horgan, G.W., and E.A. Hunter. 1993. *Introduction to REML for scientists.* Edinburgh: Scottish Agricultural Statistics Service, University of Edinburgh.

Hough, G., R.H. Sánchez, G. Garbarini de Pablo, et al. 2002. Consumer acceptability versus trained sensory panel scores of powdered milk shelf-life defects. *Journal of Dairy Science* 85: 2075–2080.

Martin, A.V. 1994. Chocolate confectionary. In *Shelf life evaluation of foods,* ed. C.M.D. Man and A.A. Jones, Chapter 11. Glasgow, UK: Blackie Academic and Professional.

Martínez, C., A. Mucci, M.J. Santa Cruz, G. Hough, and R. Sanchez. 1998. Influence of temperature, fat content and package material on the sensory shelf life of commercial mayonnaise. *Journal of Sensory Studies* 13: 331–346.

Ramírez, G., G. Hough, and A. Contarini. 2001. Influence of temperature and light exposure on sensory shelf life of a commercial sunflower oil. *Journal of Food Quality* 24: 195–204.

Restrepo, P., C. López, L. Garitta, and G. Hough. 2003. Determinación del punto de corte de café de Colombia y variación de la aceptabilidad de acuerdo a la población estudiada. Proceedings III Simposium Iberoamericano de Análisis Sensorial, Montevideo, Uruguay.

Taoukis, P., T.P. Labuza, and I.S. Saguy. 1997. Kinetics of food deterioration and shelf-life prediction. In *Handbook of food engineering practice,* ed. K. Valentas, E. Rotstein, and R.P. Singh. Boca Raton, FL: CRC Press.

# chapter 4

# Survival analysis applied to sensory shelf life

## 4.1  What is survival analysis?

Generally, survival analysis is a collection of statistical procedures for data analysis for which the outcome variable of interest is time until an event occurs (Kleinbaum 1996). The problem of analyzing time to event data arises in a number of applied fields, such as medicine, biology, public health, epidemiology, engineering, economics, and demography (Klein and Moeschberger 1997). Following is a series of examples of how time to an event of interest is considered (Gómez and Langohr 2002):

- In a clinical trial of a certain medicine, time zero is when patients are randomly allocated to treatments. Time to event is the time till cancer remission or time till a clinical indicator falls below a certain level, for example, viral count falls below 500.
- In an epidemiological study the event of interest could be the weaning of newborn baby. Time zero is the birth of the child, and time to the event is the last day the baby breastfeeds.
- Another epidemiological study could measure the time from when a person starts consuming intravenous drugs until infection with the HIV virus.
- In industrial durability tests it is of interest to know, for example, how long a car tire lasts. In this case, instead of *time to event*, *kilometers to event* is used. That is, instead of recording the time the tire is on the road, the distance in kilometers run by the car till the tire wears out is recorded.
- In a psychological study it is of interest to know, for example, the time it takes 2-year old children to learn a certain task.
- In a sociological study the event of interest could be returning to jail after having left jail for the first time. That is, the time to event would be measured from when the person was let out of jail for the first time till the person returns due to a new offence.
- In a SSL, time to event would be measured from when the product left the manufacturing plant till it was rejected by a consumer (Hough et al. 2003).

## 4.2   Censoring

Time-to-event data present themselves in different ways, which creates special problems in analyzing such data. One feature, often present in time-to-event data, is known as *censoring*, which, broadly speaking, occurs when some lifetimes are known to have occurred only within certain intervals. There are three basic categories of censoring: right-, left-, and interval-censored data.

### 4.2.1   Right-censoring

Subjects are followed till the event of interest occurs. If the event of interest does not occur during the period the subject is under study, this observation is right-censored. Continuing the above examples, this type of censoring can occur:

At the end of the study
- A cancer patient is still alive
- A tire has not worn out
- A 2-year old child is still breastfeeding
- An ex-prisoner has not returned to jail
- A consumer still accepts the sample stored for the maximum time

In the middle of a study
- A patient moves and leaves no forwarding address
- A tire bursts for extraneous reasons
- A consumer no longer wants to taste samples stored for successive times

In all these cases the event had not occurred up to a certain time, and this information is used in modeling the data.

### 4.2.2   Left-censoring

Left-censoring occurs if the subject has already undergone the event of interest before the study begins. Following are three examples of left-censored data:

- In a study on the aroma persistence of a clothes rinse, standard-sized hand towels are washed using the rinse, tumble dried, and kept in a cupboard. At different times after the application, respondents sniff a smelling strip with the aroma of the clothes rinse and are then asked if they can definitely detect the aroma on the towel. Suppose that the first test is 24 hours after the application. If a respondent cannot detect the aroma at this first test, then his data are left-censored. For this respondent, the aroma disappeared sometime between Time = 0 (application) and Time = 24 hours (first test).

- For a study to determine the distribution of the time until first marijuana use among high school students (Klein and Moeschberger 1997), the question "When did you first use marijuana?" was asked. One of the responses was, "I have used it but cannot recall just when the time was." A boy who chose this response is indicating that the event occurred prior to age at interview, but the exact age at which he started using marijuana is unknown. His data is left-censored.

- In an SSL study on mayonnaise it would not be necessary to ask consumers to taste samples with less than 2 months' storage at 25°C. If a consumer rejects a sample with 2 months' storage because she is particularly sensitive to oxidized flavor, her data are left-censored. That is, all that is known is that time to rejection for this consumer lies somewhere between Time = 0 and Time = 2 months.

## 4.2.3  Interval-censoring

Interval-censoring occurs when all that is known is that the event of interest occurred within a time interval. Following are two examples of interval-censored data:

- Longitudinal epidemiological studies are likely to have interval-censored data. For example, populations who consume intravenous drugs are highly vulnerable to infection with the HIV virus. It could be of interest to estimate the distribution of times between first intravenous drug consumption and infection with HIV. To do this, periodic blood tests are taken on willing participants. A protocol is agreed with each volunteer, for example, to perform a test every 6 months. If the HIV test was negative on June 1 but is positive on December 1, then what is known is that the subject became infected sometime between these two dates. If the subject skips a test, and the positive test is only detected on June 1 of the following year, then the infection interval is extended to 12 months.

- In SSL tests, interval-censoring is very likely to occur. In a study on the SSL of cracker-type biscuits, the maximum storage time at 20°C and 60% relative humidity is considered to be 12 months. If a reversed storage design is used (see Section 3.3.6.2), samples with different storage times are presented to consumers in a single session. To know the exact storage time at which a consumer will reject the crackers, samples would have to be taken on a daily basis. Obviously, this is not possible. Over the 12-month period, suppose that the following storage times are chosen: 0, 3, 6, 8, 10, and 12 months. If a consumer accepts the sample with 6 months' storage and rejects the sample with 8 months' storage, what is known is that her rejection time is somewhere between 6 and 8 months' storage. Her data are thus interval-censored.

In actual fact, both right-censored and left-censored data can be considered as special cases of interval-censoring. For an SSL study, with right-censored data the interval is between the last time the consumer accepted the sample and infinity, and with left-censored data the interval is between Time = 0 and the first storage time.

## 4.3    Survival and failure functions

Let T be the time of occurrence of event ε. Event ε could be death, appearance of a tumor, giving up smoking, end of itching symptoms, or a projector lamp burning out. For SSL studies, event ε is rejection of a stored product by the consumer. T is a random non-negative variable whose distribution can be characterized by the following functions:

- Survival function, S(t)
- Failure function (also referred to as cumulative distribution function), F(t)
- Probability density function, f(t)
- Hazard function, h(t)

If any of these functions is known, the others can be determined univocally. We will define the survival and failure functions.

The survival or acceptance function is the probability of an individual surviving beyond time t: S(t) = Prob(T > t) and is defined for t ≥ 0. In SSL the survival function is the probability of a consumer accepting a food product stored beyond time t and is thus referred to as the acceptance function. Figure 4.1 shows a typical survival, or acceptance, curve. Its basic properties are as follows:

- S(0) = 1: the consumer accepts the fresh product
- S(∞) = 0: the consumer rejects the product stored for prolonged periods
- S(t) is a monotonously decreasing function
- If T is continuous, S(t) is continuous and strictly decreasing

The failure or rejection function (also known as cumulative distribution function of T) is the probability of an individual failing before time t: F(t) = Prob(T ≤ t) and is defined for t ≥ 0. In SSL the rejection function is the probability of a consumer rejecting a food product stored for less than

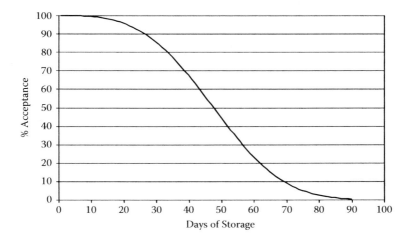

**Figure 4.1** Survival or acceptance function.

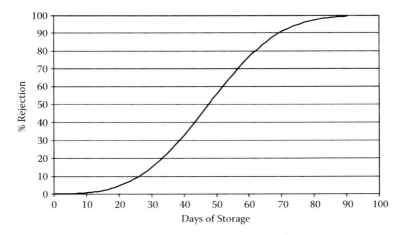

**Figure 4.2** Failure or rejection function.

time t. It can also be interpreted as the proportion of consumers who will reject a food product stored for less than time t. Figure 4.2 shows a typical failure or rejection curve.

Its basic properties are as follows:

- $F(0) = 0$: the consumer accepts the fresh product
- $F(\infty) = 1$: the consumer rejects the product stored for prolonged periods
- $F(t)$ is a monotonously increasing function
- If T is continuous, $F(t)$ is continuous and strictly increasing
- $F(t) = 1 - S(t)$

## 4.4    Shelf life centered on the product or on its interaction with the consumer?

Chapter 1 (Section 1.3) pointed out that the first step in establishing the shelf life of a product is to make sure consumers will come to no harm through eating the food during the established storage time. For some foods the nutrition aspect is crucial. Once the sanitary and nutritional hurdles have been overcome, the remaining barrier depends on the sensory properties of the product.

There are numerous examples in the food science literature where the SSL is centered on the product. For example, Fan et al. (2003) studied the use of ionizing radiation to extend the SSL of fresh-cut green onion leaves. Three assessors measured the overall quality of the onion leaves using a scale of 1 (*non-eatable*) to 9 (*excellent*). The authors decided that a value of 6 on this scale was the commercialization limit. They established that the SSL of the product was 9 days. This is what we call centering the shelf life on the product. It would be interesting to know what the consumers think about these onion leaves stored for 9 days. A very sensitive or fussy consumer would very probably find the product stored for 6 days totally unacceptable. On the other hand a less sensitive or less fussy consumer would probably be quite happy consuming onion leaves with 12 days storage. That is, from a sensory point of view, the onion leaves do not have a shelf life of their own; rather, this will depend on the interaction of the product with the consumer. The same onion leaves can be accepted by some consumers and rejected by others.

Another example was a study reported by Martínez et al. (2005) on the use of modified atmospheres to extend the shelf life of pork sausages. A six-member panel measured appearance and off-odor. For off-odor a 5-point scale was used, where 1 = *none* and 5 = *extreme*. They considered 3, defined as *small*, to be the shelf-life limit, taking this limit from a previous paper (Djenane et al. 2001) where this limit is not even mentioned, let alone justified. Here again the SSL was centered on the product, as there was no indication of what consumers would think about a *small* off-odor. That is, there is a pork sausage on a plate that has no sanitary problems. A six-member panel decides that this sausage had a *small* off-odor. This gives no indication at all on the SSL of the sausage. The response of consumers to this sausage could be:

- Reject the sausage due to the off-odor
- Not detect the off-odor as such, considering it part of the general odor to be expected in a sausage. Thus, these consumers would accept this sausage, just as they accept the fresh sample.
- Like the sausage with the off-odor more than the fresh sample, because it brings back memories of homemade sausages consumed

on a farm during childhood. These consumers would probably reject the fresh sausage and accept this off-odor one.

In a more recent article (Sirpatrawan 2009) on the SSL of rice crackers, a 10-member trained panel measured acceptability using a scale of 1 (*extremely undesirable*) to 5 (*extremely desirable*). Chapter 2 (Section 2.3) noted that using a trained panel to measure acceptability is not good practice; thus, the sensory methodology was not adequate. To determine SSL these acceptability scores were regressed versus water activity values, and considering a limiting value of 3 on the acceptability scale, critical water activities were defined. Here again, SSL is centered on the product. A cracker that had reached the critical water activity based on the arbitrary value of acceptability = 3 would be considered to have reached the end of its shelf life. Some consumers would find this cracker acceptable, and others would find it unacceptable. Also, the same consumer may find the cracker unacceptable alone and okay if accompanied with butter and jam. In Argentina it is common to dunk this type of bulky cracker in an accompanying beverage like chocolate flavored milk or coffee with milk. What happens to the critical water activity in this case?

## 4.5   Experimental data used to illustrate the methodology

To illustrate the methodology to be applied to estimate SSL using survival analysis statistics, data from a yogurt storage test will be used. A commercial, whole-fat, stirred, strawberry-flavored yogurt with strawberry pulp was used. Pots (150 g) were bought from a local distributor, all from the same batch. A reversed storage design was used (see Chapter 3, Section 3.3.6.2). The pots were kept at 4°C, and some of them were periodically placed in a 42°C oven. This particularly high temperature was chosen with the sole purpose of generating data to illustrate the methodology. Samples were stored at 42°C for the following times: 0, 4, 8, 12, 24, 36, and 48 hours. These times were chosen because a preliminary experiment showed that the flavor deteriorated quickly up to approximately 12 hours and then the deterioration slowed down. Once samples had reached the storage time at 42°C, they were refrigerated at 4°C until they were tasted; this refrigerated storage lasted between 1 and 3 days. Previous microbiological analysis (aerobic mesophiles, coliforms, yeasts, and molds) showed that the samples were fit for consumption. The ethics committee of our institute (see Chapter 2, Section 2.6.9) decided that all samples were adequate for tests on humans in the quantities to be served.

Fifty subjects who consumed stirred yogurt at least once a week were recruited from the town of Nueve de Julio (Buenos Aires, Argentina). They were presented with the seven yogurt samples (0-, 4-, 8-, 12-, 24-,

36-, and 48-hour storage time at 42°C) monadically in random order. Fifty grams of each sample was presented in a 70-ml plastic cup. Time between each sample was approximately 1 minute. Water was available for rinsing. For each sample, subjects tasted the sample and answered the question: "Would you normally consume this product? Yes or No?" It was explained that this meant: If they had bought the product to eat it, or if it was served to them at their homes, would they normally consume it or not? The tests were conducted in a sensory laboratory with individual booths with artificial daylight-type illumination, temperature control (between 22 and 24°C) and air circulation. The data obtained from the 50 consumers are in Table 4.1.

## 4.6   Censoring in shelf-life data

Table 4.2 presents the data for 5 of the 50 subjects to illustrate the interpretation given to each subject's data:

- *Subject 1* was as expected in a shelf-life study; that is, the subject accepted the samples up to a certain storage time and then consistently rejected them. The data are interval-censored because we do not know at exactly what storage time between 12 and 24 hours the consumer would start rejecting the product. Twenty-two subjects presented this type of data.
- *Subject 2* accepted all samples. Supposedly at a sufficiently long storage time (T > 48 h) the sample would be rejected and thus the data are right-censored. Eight subjects presented this type of data.
- *Subject 3* was rather inconsistent, rejecting the sample with 8 hours' storage, accepting at 12 hours' storage, and rejecting from 24 hours' storage and onward. Censoring could be interpreted in different ways. One possibility would be to consider the data as interval-censored between 4 and 8 hours; that is, ignoring the subject's answers after the first time the yogurt is rejected. Another possibility, as shown in Table 4.2, is interval-censoring between 4 and 24 hours. We consider this option as more representative of the subject's data; that is, we assign a wider uncertainty interval as to the storage time at which this subject rejects the yogurt. Eleven subjects presented this type of data.
- *Subject 4* was also rather inconsistent, with alternating *no* and *yes* answers. This subject's data were considered left-censored. Left-censoring is a special case of interval-censoring with the lower bound equal to Time = 0 (Meeker and Escobar 1998). But as the literature and statistical software distinguish it, we have also done so. The left-censoring could be considered as T ≤ 4 h or T ≤ 24 hours. As with Subject 3, a wider interval is recommended for Subject 4, as this

*Table 4.1* Consumer Acceptance/Rejection Data
for Yogurt Stored at 42°C

| Consumer | t0 | t4 | t8 | t12 | t24 | t36 | t48 |
|:---:|:---:|:---:|:---:|:---:|:---:|:---:|:---:|
| 1 | no | no | yes | yes | yes | yes | no |
| 2 | yes | yes | no | yes | no | no | no |
| 3 | yes | yes | no | yes | no | no | yes |
| 4 | yes | no | yes | yes | no | no | no |
| 5 | yes | yes | yes | yes | yes | yes | no |
| 6 | yes | yes | yes | yes | no | yes | no |
| 7 | no | yes | yes | yes | yes | yes | no |
| 8 | yes | yes | yes | yes | yes | yes | yes |
| 9 | yes | yes | yes | yes | no | no | no |
| 10 | yes | yes | yes | yes | no | yes | no |
| 11 | yes | yes | no | yes | yes | no | no |
| 12 | yes | yes | yes | yes | yes | no | no |
| 13 | yes | yes | yes | no | no | no | no |
| 14 | yes | yes | yes | yes | yes | no | no |
| 15 | yes | yes | yes | no | no | no | no |
| 16 | yes | yes | yes | yes | yes | yes | no |
| 17 | no | yes | no | yes | no | yes | yes |
| 18 | yes | yes | no | no | no | no | no |
| 19 | yes | yes | yes | yes | no | no | no |
| 20 | yes | yes | no | yes | no | yes | no |
| 21 | yes | yes | yes | no | yes | yes | yes |
| 22 | yes | yes | yes | yes | no | no | no |
| 23 | yes | yes | no | no | no | no | no |
| 24 | yes | yes | yes | yes | yes | yes | no |
| 25 | yes | yes | yes | yes | yes | no | no |
| 26 | yes | no | no | yes | no | no | no |
| 27 | yes | yes | no | no | no | no | no |
| 28 | no | yes | yes | yes | no | no | no |
| 29 | yes | yes | yes | no | yes | no | no |
| 30 | yes | yes | no | yes | no | no | no |
| 31 | yes | yes | yes | yes | yes | no | no |
| 32 | yes | yes | yes | yes | yes | no | yes |
| 33 | yes | no | no | yes | yes | no | no |
| 34 | yes | yes | yes | no | no | no | no |
| 35 | yes | yes | yes | yes | no | no | no |
| 36 | yes | yes | yes | yes | yes | yes | yes |
| 37 | yes | yes | yes | yes | yes | yes | no |

*(continued on next page)*

*Table 4.1 (continued)* Consumer Acceptance/Rejection
Data for Yogurt Stored at 42°C

| Consumer | t0 | t4 | t8 | t12 | t24 | t36 | t48 |
|---|---|---|---|---|---|---|---|
| 38 | yes | no | yes | yes | yes | no | no |
| 39 | yes | yes | yes | no | no | no | no |
| 40 | yes | no | yes | yes | yes | no | no |
| 41 | yes | yes | yes | yes | no | no | yes |
| 42 | yes | yes | yes | yes | no | yes | no |
| 43 | yes | yes | no | no | no | no | no |
| 44 | yes | yes | no | no | no | no | no |
| 45 | yes | yes | yes | yes | yes | yes | yes |
| 46 | yes | yes | yes | yes | no | yes | no |
| 47 | yes | yes | no | yes | no | no | no |
| 48 | yes | no | no | no | no | no | no |
| 49 | yes | yes | yes | yes | no | no | no |
| 50 | yes | yes | yes | yes | yes | yes | yes |

*Note:* The full data can be downloaded from the editor's Web
site: yogur.xls.

*Table 4.2* Acceptance/Rejection Data for Five Subjects Who Tasted
Yogurt Samples with Different Storage Times at 42°C

| Subject | Storage time (hours) | | | | | | | Censoring |
|---|---|---|---|---|---|---|---|---|
| | 0 | 4 | 8 | 12 | 24 | 36 | 48 | |
| 1 | yes | yes | yes | yes | no | no | no | Interval: 12–24 |
| 2 | yes | yes | yes | yes | yes | yes | yes | Right: >48 |
| 3 | yes | yes | no | yes | no | no | no | Interval: 4–24 |
| 4 | yes | no | yes | yes | no | no | no | Left: ≤24 |
| 5 | no | no | yes | yes | yes | yes | no | Not consider |

Reprinted with permission from: Hough, G., K. Langohr, G. Gómez, and A. Curia.
2003. Survival analysis applied to sensory shelf-life of foods. *Journal of Food Science*
68: 359–362.

reflects the true uncertainty of the subject's response. Five subjects
were left-censored.

- *Subject 5* rejected the fresh sample. This subject was either (a) recruited
  by mistake, that is, he did not like yogurt, or (b) he preferred the
  stored product to the fresh product, or (c) he did not understand
  the task. It would not be reasonable to consider the results of these
  subjects in establishing the shelf life of a product. For example, for
  consumers who preferred the stored to the fresh product, a company would have to produce a yogurt with a different flavor profile

*Table 4.3* Sample of Data Ready to Be Processed to Maximize the
Likelihood Function and Thus Obtain the Model's Parameters

| Subject | Low time interval | High time interval | Type of censorship |
|---------|-------------------|--------------------|--------------------|
| 1 | 12 | 24 | Interval |
| 2 | 48 | 48 | Right |
| 3 | 4 | 24 | Interval |
| 4 | 24 | 24 | Left |

rather than encourage them to consume an aged product. Four subjects presented this behavior of rejecting the fresh sample, and their
results were not considered.

Table 4.3 presents the data from the consumers of Table 4.2 in a form ready
to estimate to maximize the likelihood function and estimate the model's
parameters as explained in the following section.

## 4.7    Model to estimate the rejection function

A simple way of estimating the rejection function is to calculate the experimental percent rejection corresponding to each storage time. For example,
for a storage time of 4 hours corresponding to the yogurt data in Table 4.1,
the experimental percent rejection = 6/46 × 100 = 13%. The total number of
consumers who accepted the fresh sample was 46. Figure 4.3 presents the
results of these calculations for each storage time. This figure can be used
to obtain an approximate shelf-life value. If 50% rejection probability is
considered, estimated shelf life is 24 hours. This value is not very reliable.
Figure 4.3 shows that it covers a percent rejection between 25% and 60%.
Also, confidence intervals cannot be estimated.

The likelihood function, which is generally used to estimate the rejection function, is the joint probability of the given observations of the $n$
consumers (Klein and Moeschberger 1997):

$$L = \prod_{i \in R}\left(1 - F(r_i)\right)\prod_{i \in L} F(l_i)\prod_{i \in I}\left(F(r_i) - F(l_i)\right) \qquad (4.1)$$

where $R$ is the set of right-censored observations, $L$ the set of left-censored
observations, and $I$ the set of interval-censored observations. Equation 4.1
shows how each type of censoring contributes differently to the likelihood function.

If we can assume an appropriate distribution for the data, the use of
parametric models provides adequate estimates of the rejection function
and other values of interest. Usually, rejection times are not normally

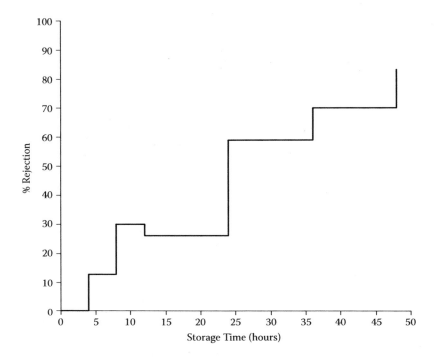

*Figure 4.3* Experimental percent rejection versus storage time for the yogurt data.

distributed; instead, their distribution is often right-skewed. Often, a log-linear model is chosen:

$$Y = \ln(T) = \mu + \sigma W$$

where $W$ is the error term distribution. That is, instead of the rejection time $T$, its logarithmic transformation is modeled. In Klein and Moeschberger (1997) or Meeker and Escobar (1998) different possible distributions for $T$ are presented, for example, the log-normal or the Weibull distribution. With the former, $W$ is the standard normal distribution; and with the latter, $W$ is the smallest extreme value distribution.

If the log-normal distribution is chosen for $T$, the rejection function is given by:

$$F(t) = \Phi \left( \frac{\ln(t) - \mu}{\sigma} \right) \tag{4.2}$$

where $\Phi(\cdot)$ is the standard normal cumulative distribution function, and $\mu$ and $\sigma$ are the model's parameters.

If the Weibull distribution is chosen, the rejection function is given by:

$$F(t) = F_{sev}\left(\frac{\ln(t) - \mu}{\sigma}\right)$$

where $F_{sev}(\cdot)$ is the rejection function of the extreme value distribution:

$$F_{sev}(w) = 1 - \exp\left(-\exp(w)\right)$$

Thus the rejection function for the Weibull distribution can be expressed as:

$$F(t) = 1 - \exp\left[-\exp\left(\frac{\ln(t) - \mu}{\sigma}\right)\right] \qquad (4.3)$$

where $\mu$ and $\sigma$ are the model's parameters. Some authors express the Weibull rejection function in another manner (Meeker and Escobar 1998; Gacula and Singh 1984):

$$F(t) = 1 - \exp\left(-\left(\frac{t}{\eta}\right)^{\beta}\right)$$

The relationship between $\beta$ and $\eta$ of this last equation and $\mu$ and $\sigma$ from Equation 4.3 is the following:

$$\sigma = 1/\beta$$

and

$$\mu = \ln(\eta)$$

Thus, either of the two expressions can be used. We prefer using Equation 4.3, as $\mu$ and $\sigma$ are the parameters calculated by the statistical packages we use.

The parameters of the loglinear model are obtained by maximizing the likelihood function (Equation 4.1). The likelihood function is a mathematical expression that describes the joint probability of obtaining the data actually observed on the subjects in the study as a function of the unknown parameters of the model considered. To estimate $\mu$ and $\sigma$ for the log-normal or the Weibull distribution, we maximize the likelihood function by substituting F(t) in Equation 4.1 by the expressions given in Equations 4.2 or 4.3, respectively.

Once the likelihood function is formed for a given model, specialized software can be used to estimate the parameters ($\mu$ and $\sigma$) that maximize the likelihood function for the given experimental data. The maximization is obtained by numerically solving the following system of equations using methods like the Newton-Raphson method (Gómez and Langohr 2002):

$$\frac{\partial \ln L(\mu, \sigma)}{\partial \mu} = 0$$

$$\frac{\partial \ln L(\mu, \sigma)}{\partial \sigma} = 0$$

For more details on likelihood functions see Klein and Moeschberger (1997) or Meeker and Escobar (1998). In practice the numerical maximization of the likelihood function is performed with specialized software such as TIBCO Spotfire S+ (TIBCO, Inc.; Seattle, WA) or the R Statistical Package (http://www.r-project.org/; accessed May 26, 2009), as will be illustrated in the following sections.

## 4.8   Calculations using the R statistical package

As mentioned in Section 4.6, one of the characteristics of SSL data is the presence of interval-censored data. Not all statistical software has the necessary procedures to deal with this type of censoring. Commercial software such as TIBCO Spotfire S+ (TIBCO, Inc., Seattle, WA) and SAS (SAS Institute, Inc., Cary, NC) have interval-censoring procedures. R, a free-access statistical package that can be downloaded from http://www.r-project.org (accessed May 26, 2009), does have a procedure for interval-censoring calculations and shall be used in the present book.

The first step in performing calculations is to have the raw data in an Excel spreadsheet, as shown in Figure 4.4 for the first 21 of the 50 consumers of Table 4.1. For this Excel spreadsheet to be compatible with the function written in R for SSL calculations, the spreadsheet should have the following characteristics:

1. The first column should indicate the consumer number. These numbers do not necessarily have to be consecutive nor start from 1, but all cells should be numbered and the column should have a text heading. In this example the heading is *consumer,* but it could be any other name.
2. The first row should contain label headings for each column, starting with the *consumer* column and following with the storage time

**Microsoft Excel - yogur.xls**

Archivo   Edición   Ver   Insertar   Formato   Herramientas   Datos   Ventana   Spotfire S+   ?   Penalty Analysis

| Consumer | T0 | T4 | T8 | T12 | T24 | T36 | T48 |
|---|---|---|---|---|---|---|---|
| 1 | no | no | si | si | si | si | no |
| 2 | si | si | no | si | no | no | no |
| 3 | si | si | no | si | no | no | si |
| 4 | si | no | si | si | no | no | no |
| 5 | si | si | si | si | si | si | no |
| 6 | si | si | si | si | no | si | no |
| 7 | no | si | si | si | si | si | no |
| 8 | si | si | si | si | si | si | si |
| 9 | si | si | si | si | no | no | no |
| 10 | si | si | si | si | no | si | no |
| 11 | si | si | no | si | si | no | no |
| 12 | si | si | si | si | si | no | no |
| 13 | si | si | si | no | no | no | no |
| 14 | si | si | si | si | si | no | no |
| 15 | si | si | si | no | no | no | no |
| 16 | si | si | si | si | si | si | no |
| 17 | no | si | no | si | no | si | si |
| 18 | si | si | no | no | no | no | no |
| 19 | si | si | si | si | no | no | no |
| 20 | si | si | no | si | no | si | no |
| 21 | si | si | si | no | si | si | si |

Hoja1 / Hoja2 / Hoja3 /

**Inicio**      R      Chapter 4.doc...   INTRUCTION...   yogur.

*Figure 4.4* Excel spreadhsheet showing raw data from a storage study of yogurt.

columns. What these columns actually say is not important. In the present example the labels are T0, T4, T8, T12, T24, T36, and T48; but they could have been time0, time4 or t1, t2, t3, and so forth. The actual storage times will be provided when using the corresponding function within R.

3. The answers of acceptance or rejection for each one of the samples by consumers can be coded as wished. In the present example the codes are *no* for rejection and *si* (*yes* in Spanish) for acceptance. The codes could be numerical, for example a 0 for rejection and a 1 for acceptance or other labels such as *rej* for rejection and *acc* for acceptance.

The data in the format shown in Figure 4.4 should be saved as a tab-delimited text file (extension *txt*) as shown in Figure 4.5. This format can be easily read by R. Once R has been installed, the instructions to analyze the yogurt data saved in the text file shown in Figure 4.5 would be as follows:

```
yogur.txt - Bloc de notas
Archivo   Edición   Formato   Ver   Ayuda
```

| Consumidor | T0 | T4 | T8 | T12 | T24 | T36 | T48 |
|---|---|---|---|---|---|---|---|
| 1 | no | no | si | si | si | si | no |
| 2 | si | si | no | si | no | no | no |
| 3 | si | si | no | si | no | no | si |
| 4 | si | no | si | si | no | no | no |
| 5 | si | si | si | si | si | si | no |
| 6 | si | si | si | si | no | si | no |
| 7 | no | si | si | si | si | si | no |
| 8 | si | si | si | si | si | si | si |
| 9 | si | si | si | si | no | no | no |
| 10 | si | si | si | si | no | si | no |
| 11 | si | si | no | si | si | no | no |
| 12 | si | si | si | si | si | no | no |
| 13 | si | si | si | no | no | no | no |
| 14 | si | si | si | si | si | no | no |
| 15 | si | si | si | no | no | no | no |
| 16 | si | si | si | si | si | si | no |
| 17 | no | si | no | si | no | si | si |
| 18 | si | si | no | no | no | no | no |
| 19 | si | si | si | si | no | no | no |
| 20 | si | si | no | si | no | si | no |
| 21 | si | si | si | no | si | si | si |
| 22 | si | si | si | si | no | no | no |
| 23 | si | si | no | no | no | no | no |
| 24 | si | si | si | si | si | si | no |
| 25 | si | si | si | si | si | no | no |
| 26 | si | no | no | si | no | no | no |
| 27 | si | si | no | no | no | no | no |
| 28 | no | si | si | si | no | no | no |
| 29 | si | si | si | no | si | no | no |
| 30 | si | si | no | si | no | no | no |
| 31 | si | si | si | si | si | no | no |
| 32 | si | si | si | si | si | no | si |
| 33 | si | no | no | si | si | no | no |
| 34 | si | si | si | no | no | no | no |
| 35 | si | si | si | si | no | no | no |

```
Inicio                    R            2 Microsof...  ▾   yogur.xls          RGui
```

*Figure 4.5* Raw data from a storage study of yogurt in shown text format.

1. Open R.
2. Go to the File Menu and change the directory to where the yogur.txt file was saved.
3. Go to the File Menu and open a New Script. This will bring up an empty window.
4. Introduce the text shown in Figure 4.6 in the New Script window. Once it has been introduced, save this window as "sslife.R"
5. What is sslife.R? It is a function for analyzing shelf-life data in a format such as shown in Table 4.1. The sslife.R function has the following format and options:

```
sslife <- function(data, tiempos = c(0, 4, 8, 12,
24, 36, 48), codiresp = c("si," "no"),model="weibu
ll,"percent=c(10,25,50))
```

```
sslife <- function(data, tiempos = c(0, 4, 8, 12, 24, 36, 48), codiresp =
c("si", "no"),model="weibull",percent=c(10,25,50))
{
   library(survival)
   totalcases <- dim(data)[1]
        casesdata <- cbind(1:totalcases, data)
        casesok <- casesdata[, 1][data[, 2] == codiresp[1]]
        numindok <- length(casesok)
        numtimes <- length(tiempos)
        id <- data[casesok, 1]
        respcod <- data[casesok, 2:dim(data)[2]]
        respnum <- matrix(rep(1, numindok * numtimes), ncol = numtimes)
        respnum[respcod == codiresp[2]] <- 0
        ti <- rep(tiempos[1], numindok)
        ts <- rep(tiempos[numtimes], numindok)
        cens <- rep("interval", numindok)
        censcod <- rep(3, numindok)
        for(i in 1:numindok) {
                if(respnum[i, numtimes] == 1) {
                        ti[i] <- tiempos[numtimes]
                        ts[i] <- tiempos[numtimes]
                        cens[i] <- "right"
                        censcod[i] <- 0
                }
                else {
                        inf <- 1
                        while(respnum[i, inf + 1] == 1) inf <- inf + 1
                        sup <- numtimes
                        while(respnum[i, sup - 1] == 0) sup <- sup - 1
                        if(inf == 1) {
                                ti[i] <- tiempos[sup]
                                ts[i] <- tiempos[sup]
                                cens[i] <- "left"
                                censcod[i] <- 2
                        }
                        else {
                                ti[i] <- tiempos[inf]
                                ts[i] <- tiempos[sup]
                        }
                }
        }
        prop<-percent/100
pp1<-data.frame(id, ti, ts, cens, censcod)
pp2<-survreg(Surv(ti,ts,censcod,type="interval")~1,dist=model)
  pp4<-predict(pp2,newdata=data.frame(1),type = "uquantile", p =
  prop, se.fit = T)
    ci3 <- cbind(pp4$fit,pp4$fit - 1.96 * pp4$se.fit,pp4$fit + 1.96 *
    pp4$se.fit)
  if (model=="weibull"|model=="lognormal"|model=="loglogistic"|model==
  "exponential") {
ci3 <- exp(ci3)
pp4$se.fit<-pp4$se.fit*ci3[,1]}
ci2<-cbind(ci3,pp4$se.fit)
mu <- c(pp2$coefficients,pp2$coefficients - 1.96 * sqrt(pp2$var[1,1]),
 pp2$coefficients + 1.96 * sqrt(pp2$var[1,1]))
```

*Figure 4.6* R-function for estimating sensory shelf life.

```
    if (model==»exponential») {
    sigma<-c(NA,NA,NA)}
    else {
    si<-exp(pp2$icoef[2])
    sigma<-c(si,exp(log(si)-1.96*sqrt(pp2$var[2,2])),exp(log(si)+1.96*
sqrt(pp2$var[2,2])))
    }
    dimnames(ci2) <- list(percent, c("Estimate","Lower ci", "Upper ci",
"Serror"))
    value<-c(«estimate»,»lower»,»upper»)
    list(censdata=pp1,musig=data.frame(value,mu,sigma),loglike=-
pp2$loglik[1],slives=ci2)
    }
```

*Figure 4.6* (continued).

- data: raw data of acceptance/rejection; no default, must be introduced.
- tiempos: food storage times; default values: c(0, 4, 8, 12, 24, 36, 48)
- codiresp: codes for consumer responses of acceptance ("si") or rejection ("no"); default values: c("si")("no").
- model: parametric model of choice (weibull, exponential, gaussian, logistic, log-normal or loglogistic); default: weibull.
- percent: percent rejection values for which estimated shelf lives are wanted; default values: c(10,25,50). For example, if estimated shelf lives are wanted for percent rejections = 10%, 20%, 30%, 40% and 50%, then percent = c(10,20,30,40,50). If a table with estimated shelf lives for a sequence of percent rejections from 10% to 50% at 1% increments is wanted, then percent = c(seq(10,50,by=1)).

6. Caveat: When text is copied from Word and pasted into R, the quotation marks are sometimes not read properly and have to be retyped in R.
7. Instructions in R are written in the R Console window after the > symbol; thus, to read the raw data the following instruction has to be written:

```
> yog <-read.table("yogurt.txt",=Header=True)
```

- <- ("less than" symbol followed by a hyphen) is supposed to symbolize an arrow and is equivalent to an equal or assignment symbol.
- Alternatively you could read the data from a directory other than the working directory; for example:

```
> yog<- read.table("C:\HOUGH\R_FILES\yogur.txt",
Header=TRUE)
```

8. Go to the File Menu, Open Source R-code: sslife.R.
9. For the yogur.txt data, use sslife with default options except for the model:

```
> resyog <- sslife(yog, model= "lognormal")
```

10. After executing the previous instruction by pressing Enter, the following message appears: "Loading required package: splines."
11. The results of the SSL analysis are in a data structure called *resyog*, which contains the censored data, the model's parameters, the log-likelihood value, and estimated storage times corresponding to different percent rejections.
12. To view the censored data:

```
> resyog$censdata
```

|    | id | ti | ts | cens     | censcod |
|----|----|----|----|----------|---------|
| 1  | 2  | 4  | 24 | interval | 3       |
| 2  | 3  | 48 | 48 | right    | 0       |
| 3  | 4  | 24 | 24 | left     | 2       |
| 4  | 5  | 36 | 48 | interval | 3       |
| ...| ...| ...| ...| ...      | ...     |
| 42 | 46 | 12 | 48 | interval | 3       |
| 43 | 47 | 4  | 24 | interval | 3       |
| 44 | 48 | 4  | 4  | left     | 2       |
| 45 | 49 | 12 | 24 | interval | 3       |
| 46 | 50 | 48 | 48 | right    | 0       |

13. The previous table corresponds with the data in Table 4.1 transformed according to the guidelines described in Section 4.6. The first column indicates the number of resulting rows. Note that there are 46 rows and not 50; this is because 4 consumers rejected the fresh sample, as can be seen in Table 4.1. The second column indicates the consumer number corresponding to Table 4.1. The third and fourth columns are the low and high time intervals corresponding to each consumer's rejection time; for right- and left-censored data the corresponding time is repeated. The fifth column indicates the type of censoring corresponding to each consumer, and the sixth column is

the code R used to interpret each type of censoring: 0, 2, and 3 for right-, left-, and interval-censored data, respectively.

14. To list the model's parameters:

```
> resyog$musig
```

|   | value | mu | sigma |
|---|-------|------|-------|
| 1 | estimate | 2.987802 | 0.9292777 |
| 2 | lower | 2.695269 | 0.7129343 |
| 3 | upper | 3.280336 | 1.2112717 |

15. The previous table shows the model's parameters—in this case the $\mu$ and $\sigma$ values corresponding to the log-normal distribution expressed in Equation 4.2. Row 1 is the mean estimate of each parameter, and rows 2 and 3 correspond to the lower and upper 95% confidence intervals, respectively. If $\mu$= 2.988 and $\sigma$= 0.929 are introduced in Equation 4.2, percent rejection versus storage time can be graphed as shown in Figure 4.7. The Excel function for the log-normal distribution is DISTR.LOG.NORM(t, $\mu$, $\sigma$).

16. To list the loglikelihood:

```
> resyog$loglike
[1]  66.7457
```

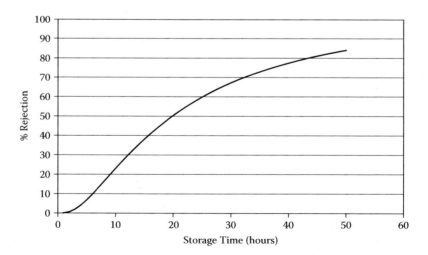

*Figure 4.7* Percent rejection versus storage time corresponding to the yogurt data (Table 4.1) for the log-normal model.

17. To compare which model best fits the data, TIBCO Spotfire S+ offers the possibility of producing a graph that compares the fit of different distributions with the experimental data, and thus a visual comparison defines which is the most adequate distribution. R does not produce this graph, and thus a way to define the best fit is to compare the loglikelihood values; the model that gives the lowest loglikelihood would be the best. In actual fact, the loglikelihood is to be used when comparing models contained one in another (Meeker and Escobar 1998); thus the criteria for choosing different models based on the loglikelihood are only approximate. In all the data we have processed, the criteria of choosing the model with the lowest loglikelihood coincided with the criteria of visual examination performed with the TIBCO Spotfire S+ software. For the present yogurt data the log-normal distribution had the lowest loglikelihood value.

18. To list the predicted shelf lives with their confidence intervals and standard errors:

```
>resyog$slives
```

| percent | estimate | lower ci | upper ci | serror |
|--------:|---------:|---------:|---------:|--------:|
| 10 | 6.030833 | 3.910189 | 9.301583 | 1.333243 |
| 25 | 10.601698 | 7.558044 | 14.871042 | 1.830425 |
| 50 | 19.842031 | 14.809504 | 26.5847 | 2.961457 |

19. In the first column of the previous table are the percent rejection values, in this case 10%, 25%, and 50%. In the second column are the estimated storage times corresponding to each percent rejection. In the third and fourth columns are the lower and upper 95% confidence limits, respectively. In the fifth column are the standard errors of the estimations.

20. Instead of listing the structures contained in *resyog* separately, they can all be listed together by simply typing:

```
> resyog
```

## 4.9    *Interpretation of shelf-life calculations*

After calculations have been performed, an SSL value has to be recommended. To do this an adequate percent rejection has to be adopted. What can be considered *adequate*? Gacula and Singh (1984) mentioned a nominal shelf-life value considering 50% rejection, and Cardelli and Labuza (2001) used this criterion in calculating the shelf life of coffee. Curia et al. (2005)

estimated SSL values of yogurt for 25% and 50% rejection probabilities. This means that if a consumer tastes a product with a storage time corresponding to 50% rejection probability, there is a 50% probability that the consumers will reject the product. This can sound too risky, yet it must be remembered that we are referring to a consumer who tastes the product at the end of its shelf life. Distribution times usually guarantee that the proportion of consumers who taste the product close to the end of its SSL is small. Of this small proportion of consumers, 50% will reject the product and 50% will accept it.

How does this 25%–50% criterion compare with values used in studies of other types of food? In determining thresholds from forced-choice data sets, the threshold value is estimated for 50% probability detection above chance (ASTM Standard E1432-04 2006; ISO Standard 13301 2002). The ISO Standard 4120 (2004) for the triangle test establishes three categories for the maximum allowable proportion of distinguishers, $p_d$:

- $p_d < 25\%$ represents small values
- $25\% < p_d < 35\%$ represents medium size values
- $p_d > 35\%$ represents large values

Thus, considering percent rejection, a probability in the range of 25% to 50% is in line with international sensory analysis standards where criteria are established to decide when the proportion of the population who can tell a difference is important.

We have had a number of clients with product on the market with a *best before* date stamped on it based on commercial experience and not on an SSL study. Their objective when conducting an SSL test was to confirm if this *best before* date was adequate. In this case, once the model's parameters μ and σ had been calculated, it was in their interest to estimate the percent rejection for their current *best before* date. If, for example, the log-normal model was chosen to model percent rejection, Equation 4.2 can be used to this purpose replacing t in the right side of the equation and calculating $F(t)$. Likewise, if the Weibull model was chosen, Equation 4.3 can be used.

An important aspect of survival analysis methodology is that experimental sensory work is relatively simple. In the above yogurt example 50 consumers each tasted seven yogurt samples with different storage times, answering *yes* or *no* to whether they would consume the samples. This information was sufficient to model the probability of consumers accepting the products with different storage times, and from the model shelf-life estimations were made. There was no necessity to have a trained sensory panel.

Another important aspect is that the information obtained from consumers by this method is directly related to their everyday eating experience. When consumers are confronted with a food product, they either accept or reject it. They do not mentally assign the product a hedonic score of 8 on a 1–9 scale and thus decide the product is acceptable, nor assign the product a score of 4 and thus decide to reject the product. Survival analysis methodology taps into direct consumer experience.

## 4.10  An additional example

SSL of fat-free stirred strawberry yogurt was studied at a storage temperature of 10°C. Yogurts were obtained from a dairy company. Bottles (1000 ml) from different batches were stored at 10°C in such a way as to have samples with different storage times ready on the same day. Storage times at 10°C were 0, 14, 28, 42, 56, 70, and 84 days. All batches were made with the same formulation and were checked to be similar to the previous batch by consensus among three expert assessors. Storage times were chosen based on a preliminary experiment. Once samples had reached the storage time at 10°C, they were refrigerated at 2°C, until they were tasted. All the measurements were made in a period not longer than a week, to guarantee no changes in the samples. To ensure that the samples were fit for consumption, the following microbiological analysis were performed on samples from the different batches and on the yogurts stored for 80 days at 10°C: coliforms, yeasts, molds, and *Staphylococcus aureus*, using standard methods of analysis (Elliott et al. 1982).

Eighty people who consumed fat-free stirred yogurt at least once a week were recruited from the city of Nueve de Julio, Buenos Aires, Argentina. Each consumer received the seven yogurt samples (corresponding to each storage time at 10°C) monadically in random order. Fifty grams of each sample were served in 70-ml plastic cups. Water was available for rinsing. For each sample subjects had to answer the question: "Would you normally consume this product: yes or no?" The data for 10 of these consumers is in Table 4.4. Of the 80 consumers, 6 rejected the fresh sample, so their data were not considered; 2 were right-censored, 18 left-censored, and 54 interval-censored. As mentioned in Section 4.8, to process the data with R, the data must be stored in a tab-delimited text file. Suppose we store this data in "strawberry.txt."

1. In the 'R Console' window write the following instruction:

```
> straw <-read.table("strawberry.txt",header=TRUE)
```

Alternatively you could read the data from a directory other than the working directory, for example:

```
> yog<- read.table("C:\HOUGH\R_FILES\strawberry.
txt",header=TRUE)
```

*Table 4.4* Acceptance/Rejection Data for 10 Consumers
Who Evaluated Fat-Free Stirred Yogurt Stored at 10°C

| Consumer | t0 | t14 | t28 | t42 | t56 | t70 | t84 |
|----------|----|-----|-----|-----|-----|-----|-----|
| 1  | 0 | 0 | 1 | 0 | 1 | 1 | 1 |
| 2  | 0 | 0 | 1 | 0 | 0 | 1 | 1 |
| 3  | 0 | 0 | 1 | 0 | 0 | 1 | 1 |
| 4  | 0 | 1 | 0 | 1 | 0 | 1 | 1 |
| 5  | 0 | 0 | 1 | 1 | 1 | 1 | 1 |
| 6  | 0 | 1 | 1 | 0 | 0 | 1 | 1 |
| 7  | 0 | 1 | 0 | 1 | 0 | 1 | 1 |
| 8  | 0 | 0 | 0 | 0 | 1 | 1 | 1 |
| 9  | 0 | 0 | 1 | 0 | 0 | 1 | 1 |
| 10 | 0 | 0 | 1 | 0 | 1 | 1 | 1 |

*Note:* Acceptance is symbolized by 0 and rejection by 1. The full
data can be downloaded from the editor's Web site:
strawberry.xls.

2. Go to the File Menu, Open Source R-code: sslife.R.
3. For the strawberry data, sslife has to receive the data (straw), the
   time points (c(0,14,28,42,56,70,84)), the response codes (c("0","1")) and
   the model ("log-normal"):

```
> resstraw<-sslife(straw,c(0,14,28,42,56,70,84),
c("0","1"),model="lognormal")
```

4. The results of the SSL analysis are in a data structure called *resstraw*,
   which contains the censored data, the model's parameters, the loglikelihood value, and estimated storage times corresponding to different percent rejections (in this case we adopted the default values
   of 10%, 25%, and 50%). The first step will be to choose the most adequate model. To do this type:

```
> resstraw$loglike
[1] 92.78819
```

5. Repeat Steps 3 and 4 for the other possible models: exponential,
   Gaussian, logistic, loglogistic, and Weibull. The resulting loglikelihood values are in Table 4.5. The logistic distribution gives a slightly
   lower loglikelihood value than the Gaussian (or normal) distribution. The shape of the logistic distribution is very similar to that of
   the normal distribution; in fact, it would require an extremely large
   number of observations to assess whether data come from a normal
   or logistic distribution (Meeker and Escobar 1998). The loglikelihood

*Table 4.5* Loglikelihood Values for Different Models Corresponding to the Fat-Free Strawberry Yogurt Data

| Model | Loglikelihood |
|---|---|
| Logistic | 83.6 |
| Gaussian | 84.1 |
| Weibull | 84.9 |
| Loglogistic | 90.3 |
| Log-normal | 92.8 |
| Exponential | 116 |

values for these two distributions have always been very similar for SSL data we have processed. Thus, the logistic function would rarely need to be considered. The hazard function corresponding to the exponential function is constant, that is, it would correspond to elements or populations that do not age. Certain electronic components behave in this manner, but not food. If a yogurt has been stored for 10 days at 10°C and was accepted by the consumer, the probability of this consumer rejecting the yogurt on the following day is low. However, if a consumer accepts a yogurt stored for 40 days, the probability of this consumer rejecting the yogurt on the following day is high. This occurs because the product has aged. Thus the exponential function would not be adequate for SSL studies of food products. In all cases that we have tested it, its loglikelihood has been higher than other models. Going back to Table 4.5, the choice would be between the normal and Weibull distributions. We shall choose the normal distribution.

6. Calculations for the normal or Gaussian distribution are performed with the following instruction in R:

```
> resstraw<-sslife(straw,c(0,14,28,42,56,70,84),
c("0","1"),model="gaussian")
```

7. To obtain the censored data:

```
> resstraw$censdata
```

|  | id | ti | ts | cens | censcod |
|---|---|---|---|---|---|
| 1 | 1 | 14 | 56 | interval | 3 |
| 2 | 3 | 14 | 70 | interval | 3 |
| 3 | 4 | 14 | 70 | interval | 3 |
| 4 | 5 | 70 | 70 | left | 2 |
| ... | ... | ... | ... | ... | ... |
| 70 | 76 | 42 | 56 | interval | 3 |
| 71 | 77 | 42 | 42 | left | 2 |
| 72 | 78 | 28 | 42 | interval | 3 |
| 73 | 79 | 42 | 56 | interval | 3 |
| 74 | 80 | 28 | 42 | interval | 3 |

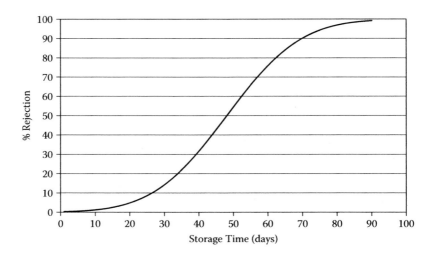

*Figure 4.8* Percent rejection versus storage time corresponding to the fat-free stirred strawberry yogurt data for the normal model.

8. To list the models parameters:

> resstraw$musig

|   | value    | mu       | sigma    |
|---|----------|----------|----------|
| 1 | estimate | 48.03002 | 16.84253 |
| 2 | lower    | 43.53564 | 13.74913 |
| 3 | upper    | 52.5244  | 20.63189 |

9. With the values of the previous table the percent rejection versus storage time curve can be drawn as shown in Figure 4.8. The Excel function for the cumulative normal distribution is =DISTR. NORM(t,$\mu$,$\sigma$,TRUE)*100.
10. To list the predicted shelf lives with their confidence intervals and standard errors:

>resstraw$slives

| percent | estimate | lower ci | upper ci | serror   |
|---------|----------|----------|----------|----------|
| 10      | 26.44545 | 19.66408 | 33.22683 | 3.459885 |
| 25      | 36.66991 | 31.28582 | 42.054   | 2.746984 |
| 50      | 48.03002 | 43.53564 | 52.5244  | 2.293051 |

11. In this study the company had been using a *best before* date corresponding to 35 days' shelf life. The previous table shows that this corresponds to approximately 25% rejection probability, which was considered acceptable, and thus the company maintained the 35 days' shelf life with renewed confidence.

12. As shown in Table 4.5 the normal and Weibull distributions had similar loglikelihood values. What would have been the results if the Weibull distribution had been chosen? Calculations can be performed with the following instruction in R:

```
> resweib<-sslife(straw,c(0,14,28,42,56,70,84),
c("0","1"),model="weibull")
```

13. To list the model's parameters:

```
> resweib$musig
```

|   | value | mu | sigma |
|---|-------|-----|-------|
| 1 | estimate | 3.981363 | 0.3098097 |
| 2 | lower | 3.896379 | 0.2443884 |
| 3 | upper | 4.066348 | 0.392744 |

14. With the values of the previous table the percent rejection versus storage time curve can be drawn using Equation 4.3. This will not be done as the curve virtually superimposes the normal distribution curve shown in Figure 4.8.

15. To list the predicted shelf lives with their confidence intervals and standard errors:

```
>resweib$slives
```

| percent | estimate | lower ci | upper ci | serror |
|---------|----------|----------|----------|--------|
| 10 | 26.68703 | 21.53232 | 33.07574 | 2.922265 |
| 25 | 36.42923 | 31.41094 | 42.24926 | 2.754782 |
| 50 | 47.8377 | 43.32459 | 52.82093 | 2.418575 |

16. These shelf-life values estimated using the Weibull distribution are very similar to those obtained using the normal distribution (see Step 10, above). Thus in practice, when the loglikelihood values are similar, final results do not differ. In these cases we tend to prefer the Weibull distribution as it is very flexible and right-skewed

and thus particularly appropriate for modeling survival data and it has been used previously in food shelf-life modeling (Cardelli and Labuza 2001; Duyvesteyn et al. 2001; Hough et al. 1999)

## 4.11   Should consumers be informed?

For SSL studies based on survival analysis statistics, samples with different storage times are presented to consumers. A practical question is whether consumers should be informed that they are evaluating samples with different storage times. Arguments *against* telling them would be as follows:

- If they are told they will pay special attention to samples, while in a natural setting they would not. The effect of this would be a shorter *best before* date than really necessary.
- They feel they are expected to find a sample that should be rejected.

Consider the case of a person who decides to open a carton of UHT milk on June 5, 2009. Supposing the *best before* date on the carton is October 5, 2009. In this case the person will be predisposed to thinking the milk is okay and will probably pay little attention to its sensory properties. If, however, the *best before* date is June 6, 2009 this will strike an alarm that the product may be of doubtful quality and now the person will pay attention to its sensory properties. With milk, the person was probably going to prepare a pudding or white sauce where any slight off-flavor would go unnoticed, but knowing the product is close to its *best before* date will lead the person to taste the milk on its own to see whether it is okay. It is for these consumers who taste the product close to its *best before* date and who thus pay special attention to its sensory properties that we are going to the trouble of estimating SSL. It is our policy to inform consumers that they are going to be evaluating samples with different storage times.

## 4.12   Is there a way to deal with totally new products?

In the experiments described in Sections 4.5 and 4.10 dealing with SSL of yogurts, regular consumers of the yogurts were recruited. When presented with the test samples, it was implied that they compare these to their internal reference of an acceptable product, and based on this comparison would tick either the *accept* or the *reject* box.

But what happens when the objective is to measure the SSL of a product that is new to the market? On one occasion we were asked to determine the SSL of canola oil, which was a product totally new to the

Argentine market. Canola oil has a distinct flavor. If an Argentine consumer were presented with a sample of stored canola oil and found that it had an odd flavor, she would not know whether this odd flavor was due to prolonged storage or was the typical flavor of canola oil. Because there was no internal reference, we presented consumers with a sample labeled *canola oil* and informed them that this was a fresh sample of the product with acceptable sensory properties. After this they were asked to evaluate the stored samples and tick one of two boxes labeled with *accept* or *reject*. This solution is not very satisfactory. In all probability consumers performed a discrimination task rather than an acceptability evaluation. That is, if they found a sensory difference between a stored sample and the "acceptable" control, they would most probably tick the *reject* box. For totally new products the safest approach is to conduct a discrimination test as described in Section 2.7.1.2. The resulting *best before* date would probably be too conservative. Once the product has been established on the market a consumer-based SSL determination can be made.

## References

ASTM Standard E1432–04. 2006. Standard practice for defining and calculating individual and group sensory thresholds from forced-choice data sets of intermediate size. West Conshohocken, PA: American Society for Testing and Materials.

Cardelli, C., and T.P. Labuza. 2001. Application of Weibull hazard analysis to the determination of the shelf life of roasted and ground coffee. *Lebensmittel-Wissenschaft und-Technologie* 34: 273–278.

Curia, A., M. Aguerrido, K. Langohr, and G. Hough. 2005. Survival analysis applied to sensory shelf life of yogurts. I: Argentine formulations. *Journal of Food Science* 70: S442–445.

Djenane, D., A. Sánchez-Escalante, J.A. Beltrán, and P. Roncalés. 2001. Extension of the retail display life of fresh beef packaged in modified atmosphere by varying lighting conditions. *Journal of Food Science* 66: 181–186.

Duyvesteyn, W.S., E. Shimoni, and T.P. Labuza. 2001. Determination of the end of shelf-life for milk using Weibull hazard method. *Lebensmittel-Wissenschaft und-Technologie* 34: 143–148.

Elliott R.P., D.S. Clark, K.H. Lewis, H. Lundbeck, J.C. Olson, and B. Simonsen. 1982. *Microorganismos de los alimentos. Técnicas de análisis microbiológico.* Volumen I. Zaragoza, Spain: Editorial Acribia.

Fan, X., B.A. Niemira, and K.J.B. Sokorai. 2003. Use of ionizing radiation to improve the sensory and microbial quality of fresh-cut green onion leaves. *Journal of Food Science* 68: 1478–1463.

Gacula M.C., and J. Singh. 1984. *Statistical methods in food and consumer research,* Chapter 11. Orlando, FL: Academic Press Inc.

Gómez, G., and K. Langohr. 2002. Análisis de supervivencia. Barcelona, Spain: Universitat Politécnica de Catalunya.

Hough, G., K. Langohr, G. Gómez, and A. Curia. 2003. Survival analysis applied to sensory shelf-life of foods. *Journal of Food Science* 68: 359–362.

Hough, G., M.L. Puglieso, R. Sánchez, and O. Mendes da Silva. 1999. Sensory and microbiological shelf-life of a commercial Ricotta cheese. *Journal of Dairy Science* 82: 454–459.

ISO Standard 13301. 2002. Sensory analysis—Methodology—General guidance for measuring odour, flavour and taste detection thresholds by a three-alternative forced-choice (3-AFC) procedure. Geneva: International Standard Organization.

ISO Standard 4120. 2004. Sensory analysis—Methodology—Triangle test. Geneva: International Standard Organization.

Klein, J.P., and M.L. Moeschberger. 1997. *Survival analysis, techniques for censored and truncated data.* New York: Springer-Verlag Inc.

Kleinbaum, D.G. 1996. *Survival analysis: A self-learning text.* New York: Springer-Verlag Inc.

Martínez, L., D. Djenane, I. Cilla, J.A. Beltrán, and P. Roncalés. 2005. Effect of different concentrations of carbon dioxide and low concentration of carbon monoxide on the shelf-life of fresh pork sausages packaged in modified atmosphere. *Meat Science* 71: 563–570.

Meeker, W.Q., and L.A. Escobar. 1998. *Statistical methods for reliability data.* New York: John Wiley & Sons.

Sirpatrawan, U. 2009. Shelf-life simulation of packaged rice crackers. *Journal of Food Quality* 32: 224–239.

*chapter 5*

# Survival analysis continued
## Number of consumers, current status data, and covariates

## 5.1  Number of consumers

Survival analysis has been applied to estimating the shelf life of foods based on consumers' accepting or rejecting samples stored at different times (Hough et al. 2003; Gámbaro et al. 2004; Curia et al. 2005; Salvador et al. 2005; Hough et al. 2006a). The number of consumers used for these estimations has varied between 50 and 80. Although shelf-life confidence intervals have been reasonable, there are no systematic criteria in the choice of the number of consumers.

Numbers of consumers necessary for sensory acceptability tests using hedonic scales have been calculated by Hough et al. (2006b). Meeker and Escobar (1998) give different results on sample size determination for life tests. However, their results are only appropriate for right-censored data and cannot be applied in consumers' accepting or rejecting frameworks where interval-censored observations are present. As an alternative, also discussed in Meeker and Escobar (1986), simulation has proven to be a powerful tool for planning survival analysis experiments.

Hough et al. (2007) published an article describing a simulation study to estimate the number of consumers necessary for shelf-life estimations based on survival analysis statistics. Their simulation method consisted of the following steps: (a) specification of a model (Weibull) and parameter values of the lifetime distribution ($\sigma$ values were taken from previous studies); (b) simulation data from the proposed life test with a specific sample size; (c) analysis of the data and assessment of the precision of estimates; and (d) repetition of the simulation process with different sample sizes and different parameter values. This process provided information on sample size requirements to achieve the desired precision for different parameter settings.

The results (Hough et al. 2007) were presented in the form of operational curves: Type II $\beta$ error versus $\Delta t$; where $\Delta t$ is the difference we are willing to accept between the true shelf life and the estimated shelf life. These curves were presented for values of the σ Weibull parameters of 0.17, 0.44, and 0.71; and for supposed shelf-life values of 2, 3, and 4 in an adimensional 0–6 time scale.

If a researcher has no previous knowledge of the σ value of his system, an intermediate value of σ = 0.44 will have to be assumed. Also he can suppose that the shelf life will be close to the middle of the times studied, that is, t = 3 in the adimensional time system used by Hough et al. (2007). A reasonable Δt would be 0.5 in the 0–6 adimensional time scale. A Type I α error = 0.05 was considered, and choosing a Type II β error = 0.2, which corresponds to a power = 0.8, allows reading a value of approximately N = 120 from the operational curves shown in Figure 5.1. This would be a recommended number of consumers for sensory shelf-life (SSL) studies based on survival analysis statistics. In most of the examples presented in this book, lower numbers of consumers are reported. This is because chronologically the estimation of the necessary number of consumers was done after having worked on the examples.

## 5.2   Current status data

### 5.2.1   Introduction

In the examples presented in the previous chapter, each consumer tasted the whole set of samples corresponding to the different storage times. Because reversed storage designs were used (see Section 3.3.6.2), it was possible for each consumer to taste all samples in a single session in random order. The reversed storage design is not always possible or convenient, for example, when there is difficulty in storing samples in conditions where sensory changes do not occur or occur very slowly. As indicated in Section 3.3.6.2, if a storage condition where the sample remains unchanged cannot be obtained, an alternative for a reversed storage design would be to introduce samples from different production batches over the designed storage time. This can be done if there is certainty that the variations between batches are sufficiently small. Otherwise, the batch variations will be confused with the storage time variations. In the cases in which the reversed storage design is not possible, the basic design can be used (Section 3.3.6.1).

For the basic design, the same consumers would have to be assembled for each one of the storage times; this procedure can be complicated and costly. Also, for products with relatively long storage times, there can be a drift in consumer habits. For example, a consumer can be recruited at the beginning of a study as a regular consumer of a certain mayonnaise

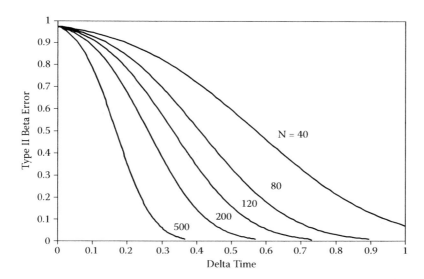

*Figure 5.1* Operational curves of Type II β error versus Δt for an average value of the σ Weibull parameter = 0.44, a shelf-life value corresponding to the middle of the time scale, and a Type I α value of 0.05. (Reprinted with permission from Hough, G., Calle, M.L., Serrat, C., and Curia, A. 2007. Number of consumers necessary for shelf-life estimations based on survival analysis statistics. *Food Quality and Preference* 18: 771–775.)

brand, and after 7 months, when the SSL study is finishing, he may have stopped using mayonnaise or changed brands. This could lead to a rejection of a stored sample due to a change of consumer habits and not due to product deterioration. Also consumers get to know they are participating in a study, for example, a mayonnaise study, and very probably start paying attention to their everyday consumption of the product, thus gaining a certain degree of expertise, which is undesirable in a consumer study. Another drawback to the basic design is that trained assessors and/or consumers can become aware that they are participating in a SSL study. This can lead to biased results. If a consumer rejects today's sample, when she is asked to come in again to taste another sample in 20 days' time, expectation will lead her to reject that sample also.

An alternative methodology that avoids the above drawbacks of the basic and reversed storage designs is to have each consumer taste a single sample corresponding to a single storage time. The data generated is given the name of current-status (Shiboski 1998) or quantal-response data (Meeker and Escobar 1998).

## 5.2.2 Experimental data

To illustrate the methodology, data were generated by storing cracker-type biscuits placed on a rack in a closed vessel, with water at the base. Thus the crackers were submitted to 90% relative humidity for 0.3, 1, 2, 4, 7, 9, and 14 hours. Once they had reached the corresponding storage time they were stored in double polyethylene bags to preserve their moisture content till served to consumers. For each one of the 7 storage times a group of 50 consumers was recruited, aged between 15 and 18 years. The total number of participating consumers was 350: 50 consumers for each of 7 storage times. Each person was a regular consumer of cracker type biscuits. The consumers tasted a single sample corresponding to their assigned storage time. They received two crackers of the sample and were informed that these crackers had been stored for a certain time (see Section 4.11 where informing consumers they are receiving stored samples is discussed). They were asked if they would normally accept or reject the biscuits presented to them.

If a consumer tastes a cracker with 4 hours' storage time, there are two alternatives:

1. He accepts it: This means that his data are right-censored. That is, this consumer's rejection time is > 4 hours.
2. He rejects it: This means that his data are left-censored. That is, this consumer's rejection time is < 4 hours.

The data generated by this experiment for 15 of the 350 consumers is shown in Table 5.1. For this data to be processed by the R statistical package, it has to be in the form shown in Table 5.2.

## 5.2.3 Model and data analysis

The survival analysis model to be applied to current status data is basically the same model as described in Section 4.7. In that case there were three types of censoring: left, interval, and right. For current status data there is no interval-censoring and thus the likelihood function is:

$$L = \prod_{i \in R} \left(1 - F(r_i)\right) \prod_{i \in L} F(l_i)$$

where $R$ is the set of right-censored observations and $L$ is the set of left-censored observations. The equation above shows how each type of censoring contributes differently to the likelihood function.

*Table 5.1* Current Status Data Obtained from Cracker-Type Biscuits Stored at 90% Relative Humidity

| Consumer | Storage time (hours) | Answer | Censoring |
|---|---|---|---|
| 1 | 0.3 | accept | right |
| 2 | 0.3 | accept | right |
| 3 | 0.3 | accept | right |
| 4 | 0.3 | accept | right |
| 5 | 0.3 | accept | right |
| ... | ... | ... | ... |
| 151 | 4 | reject | left |
| 152 | 4 | reject | left |
| 153 | 4 | accept | right |
| 154 | 4 | accept | right |
| 155 | 4 | accept | right |
| ... | ... | ... | ... |
| 346 | 14 | reject | left |
| 347 | 14 | reject | left |
| 348 | 14 | accept | right |
| 349 | 14 | reject | left |
| 350 | 14 | reject | left |

Statistical software used to analyze current status data is the same as that used to analyze interval-censored data. As mentioned in Section 4.8, not all statistical software has the necessary procedures to deal with this type of censoring. R, a free-access statistical package (see http://www.r-project.org/, accessed May 26, 2009) does have a procedure for interval-censoring calculations and is used in this analysis of current status data.

The first step in performing calculations is to have the raw data in an Excel spreadsheet in a format as shown in Table 5.2. For this Excel spreadsheet to be compatible with the function written in R for SSL calculations, it should have the following characteristics:

1. The first column should indicate the consumer number. These numbers do not necessarily have to be consecutive nor start from 1, but each cell should be numbered and the column should have a text heading. In this example the heading is *cons*, but it could be any other name.
2. The second and third columns represent the storage times corresponding to the low and high times of the interval, respectively.

*Table 5.2* Current Status Data Obtained from Cracker-Type Biscuits Stored at 90% Relative Humidity

| cons[a] | tlow | thigh | censor | censcod |
|---------|------|-------|--------|---------|
| 1 | 0.3 | 0.3 | right | 0 |
| 2 | 0.3 | 0.3 | right | 0 |
| 3 | 0.3 | 0.3 | right | 0 |
| 4 | 0.3 | 0.3 | right | 0 |
| 5 | 0.3 | 0.3 | right | 0 |
| ... | ... | ... | ... | |
| 151 | 4 | 4 | left | 2 |
| 152 | 4 | 4 | left | 2 |
| 153 | 4 | 4 | right | 0 |
| 154 | 4 | 4 | right | 0 |
| 155 | 4 | 4 | right | 0 |
| ... | ... | ... | ... | |
| 346 | 14 | 14 | left | 2 |
| 347 | 14 | 14 | left | 2 |
| 348 | 14 | 14 | right | 0 |
| 349 | 14 | 14 | left | 2 |
| 350 | 14 | 14 | left | 2 |

*Note:* Data are in the format necessary to be analyzed by the sslcsd (see Figure 5.2) function of the R-statistical package. The full data can be downloaded from the editor's Web site: cracker.xls.

[a] Column headings: cons, consumer; tlow, low time interval (hours); thigh, high time interval (hours); censor, type of censorship; censcod, censorship code.

These do not have to have the same names as in Table 5.2, but they must have text headings. For current status data the numbers in both columns are the same. Another option would be to label the *thigh* entries with *NA* (not available), which is the missing value symbol used by *R*.

3. The fourth and fifth columns represent the type of censoring that corresponds to each of the consumers, represented by a text (*Right* or *Left*) and a code (0 and 2), respectively. The headings of these columns must be present, although they do not have to have the same names as in Table 5.2.

The data in the format shown in Table 5.2 should be saved as a tab-delimited text file. This format can be easily read by R. The instructions to analyze the cracker data saved in the text file cracker.txt are as follows:

```
sslcsd <- function(daten,model="weibull",percent=c(10,25,50))
{
    library(survival)
    prop<-percent/100
    pp1<-daten
    value<-c("estimate","lower","upper")
    pp2<-survreg(Surv(daten[,2],daten[,3],daten[,5],type="interval")~1,
data=daten,dist=model)
      pp4<-predict(pp2,newdata=data.frame(1),type = "uquantile",
p = prop, se.fit = T)
      ci3 <- cbind(pp4$fit,pp4$fit - 1.96 * pp4$se.fit,pp4$fit + 1.96 *
pp4$se.fit)
      if (model=="weibull"|model=="lognormal"|model=="loglogistic"|
model=="exponential") {
      ci3 <- exp(ci3)
      pp4$se.fit<-pp4$se.fit*ci3[,1]}
      ci2<-cbind(ci3,pp4$se.fit)
      mu <- c(pp2$coefficients,pp2$coefficients - 1.96 *
sqrt(pp2$var[1,1]),pp2$coefficients + 1.96 * sqrt(pp2$var[1,1]))
      si<-exp(pp2$icoef[2])
      if (model=="exponential") {
        sigma<-c(NA,NA,NA) }
      else {
        si<-exp(pp2$icoef[2])
        sigma<-c(si,exp(log(si)-1.96*sqrt(pp2$var[2,2])),exp(log(si)+1.96
*sqrt(pp2$var[2,2])))
        }

    dimnames(ci2) <- list(percent, c("Estimate","Lower ci",
"Upper ci","Serror"))

    list(censdata=pp1,musig=data.frame(value,mu,sigma),
loglike=-pp2$loglik[1],slives=ci2)

}
```

*Figure 5.2* R-function for estimating sensory shelf life from current status data.

1. Open R.
2. Go to the File Menu and change the directory to that where the cracker.txt file was saved.
3. Go to the File Menu and open a New Script. This will bring up an empty window.
4. Introduce the text shown in Figure 5.2 in the New Script window. Once it has been introduced, save this window as *sslcsd.R*.
5. What is *sslcsd.R*? It is a function that will allow analyzing shelf-life data in a format such as shown in Table 5.2. The sslcsd.R function has the following format and options:

```
sslcsd <- function(daten, model="weibull",percent=
c(10,25,50))
```

- daten: censored data.
- model: parametric model of choice (weibull, exponential, gaussian, logistic, log-normal, or loglogistic).
- percent: percent rejection values for which I want estimated shelf lives. For example, if I want estimated shelf lives for percent rejections = 10%, 20%, 30%, 40%, and 50%, then percent = c(10,20,30,40,50). If I want a table with estimated shelf lives for a sequence of percent rejections from 10% to 50% at 1% increments, then percent = c(seq(10,50,by=1)).

6. Caveat: When text is copied from Word and pasted into R, the quotation symbols are sometimes not read properly and have to be retyped in R.
7. Instructions in R are written in the R Console window after the > symbol; thus, to read the raw data the following instruction has to be written:

```
> crac <-read.table("cracker.txt",header=TRUE)
```

- <- (*less than* symbol followed by a hyphen) is supposed to symbolize an arrow and is equivalent to an equal or assignment symbol.
- Alternatively you could read the data from a directory other than the working directory; for example:

```
> crac<- read.table("C:\HOUGH\R_FILES\cracker.
txt",header=TRUE)
```

8. Go to the File Menu, Open Source R-code: sslcsd.R.
9. For the cracker.txt data, use sslcsd with default options:

```
> rescrac <- sslcsd(crac)
```

10. After executing the previous instruction by pressing Enter, the following message appears: Loading required package: splines.
11. The results of the SSL analysis are in a data structure called *rescrac*, which contains the censored data, the model's parameters, the log-likelihood value, and estimated storage times corresponding to different percent rejections. Each one of these can be displayed separately by typing: rescrac$censdata, rescrac$musig, rescrac$loglike, and rescrac$slives; respectively.
12. Or they can all be displayed together by typing:

```
> rescrac
```

`$censdata`

|     | id  | ti  | ts  | cens  | censcod |
|----:|----:|----:|----:|-------|--------:|
| 1   | 1   | 0.3 | 0.3 | Right | 0       |
| 2   | 2   | 0.3 | 0.3 | Right | 0       |
| 3   | 3   | 0.3 | 0.3 | Right | 0       |
| 4   | 4   | 0.3 | 0.3 | Right | 0       |
| ... | ... | ... | ... | ...   | ...     |
| 346 | 346 | 14  | 14  | Left  | 2       |
| 347 | 347 | 14  | 14  | Left  | 2       |
| 348 | 348 | 14  | 14  | Right | 0       |
| 349 | 349 | 14  | 14  | Left  | 2       |
| 350 | 350 | 14  | 14  | Left  | 2       |

`$musig`

|   | value    | mu       | sigma     |
|---|----------|----------|-----------|
| 1 | estimate | 2.244178 | 0.9313116 |
| 2 | Lower    | 2.059836 | 0.7430001 |
| 3 | Upper    | 2.42852  | 1.1673502 |

`$loglike`

`[1] 168.8685`

`$slives`

| percent | estimate | lower ci  | upper ci | serror    |
|--------:|----------|-----------|----------|-----------|
| 10      | 1.159959 | 0.7500946 | 1.793781 | 0.2579968 |
| 25      | 2.956061 | 2.2908368 | 3.814456 | 0.3844998 |
| 50      | 6.704912 | 5.6577007 | 7.945957 | 0.5809425 |

The Weibull model was chosen because it had the lowest loglikelihood value in comparison to the other possible models (see Step 5 above). For example with:

```
> rescrac<-sslcsd(crac,"lognormal")
rescrac$loglike
[1] 170.5117
```

which is higher than the loglikelihood value obtained with the Weibull distribution.

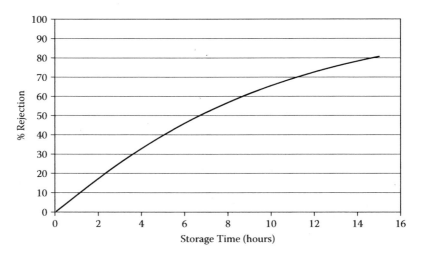

**Figure 5.3** Percent rejection versus storage time corresponding to cracker-type biscuits stored at 90% relative humidity for the Weibull model.

With $\mu = 2.244$ and $\sigma = 0.931$ (see Step 12 above) the rejection probability for the Weibull distribution given by Equation 4.3 can be calculated and graphed as shown in Figure 5.3.

## 5.2.4   Conclusions on current status data

In interpreting data from survival analysis methodology (see Section 4.9) it was pointed out that experimental sensory work is relatively simple. For current status data each consumer evaluates a single sample, stating whether he accepts or rejects the sample. This is very simple data to obtain. It could be argued that a relatively large number of consumers have to be recruited (in the cracker biscuit example, 350), but this is made easy due to the simple and quick task that each consumer has to perform.

In the introductory Section 5.2.1 current status data was presented as an alternative when reversed storage designs were inappropriate and also as a means of avoiding expectation errors in basic designs. If a reversed storage design (see Section 3.3.6.2) was to be used for ready-to-eat lettuce stored at 4°C, a first batch of lettuce would be placed at 4°C and this would correspond to the longest storage time. A second batch, harvested 3 days later, would be placed at 4°C, and this would correspond to the second longest storage time. This process would continue until all storage times have been completed. This system has the advantage of being able to measure all samples on the same day but has the disadvantage of having storage times and batches confused. For lettuce and most other vegetables this would be the case as batches are variable from one harvesting time to another.

The other reverse storage design uses a single batch of product, but as the test items are removed from storage they are frozen so they can all be evaluated together at a future date. Lettuce and other vegetables suffer considerably if frozen. The other alternative is to have the same consumers perform repeated tests for each one of the storage times. The disadvantages of this procedure were discussed in Section 5.2.1. Due to these difficulties Araneda et al. (2008) applied current status methodology to estimate the sensory shelf life of ready-to-eat lettuce (*Lactuca sativa* var. capitata cv. Alpha). For each of six storage times, 50–52 consumers answered if they would normally consume the presented sample—yes or no. Results were satisfactory and showed that this methodology can be applied when necessary. The Weibull model was found adequate to model the data. Estimated shelf lives ± 95% confidence intervals were 11.3 ± 1.2 and 15.5 ± 0.9 days for a 25% and 50% consumer rejection probability, respectively.

Another potential use for current status data methodology is to study the SSL of products that are best evaluated in a home setting. In Argentina and other South American countries (Brazil, Paraguay, and Uruguay) a very popular beverage is yerba maté (*Ilex paraguariensis* Saint Hilaire). The most traditional way of consuming this beverage is using a maté (Santa Cruz et al. 2001), a flask of approximately 200 ml made of different materials, for example, gourd (*Lagenaria vulgaris*), wood, ceramic, disposable plastic, and others. Approximately 30 g of yerba maté are placed in those flasks, and approximately 40 ml of hot water (between 70 and 85°C) is poured over it. The resulting beverage is sucked through a tube that has a sieve at one end. Repeated portions (between 8 and 10) of hot water are poured over the same yerba maté and sucked. There are many different customs regarding the consumption of yerba mate, for example, using water at different temperatures, adding different quantities of sugar, not drinking the first maté because it is too strong, adding orange or lemon peels, and so forth. For an SSL study, to have consumers come into a central location and customize the maté preparations would be very cumbersome. Another possibility would be to give the consumers the 6–7 samples, each corresponding to a different storage time, and ask them to evaluate the samples in their homes, preparing the maté as they usually do. Logistics and explanations to consumers have to be carefully considered to avoid confusions. Current status data methodology would be a good solution to having consumers evaluate the samples with different storage times. Simply give each consumer a single sample to take home, consume the product as customary and then retrieve the *accept* or *reject* response. There are different methods of obtaining the answers: phone, text message, e-mail, or a visit.

Another case where current status data methodology would be convenient is for an alcoholic beverage such as beer. Having consumers participate in a central location test where they have to taste 6 or 7 beer samples would not be convenient. With this number of samples the carry-over effect could be important and the alcohol intake could raise safety issues. It would be better for each consumer to receive a single bottle of beer to take home.

## 5.3   Introducing covariates in the model

### 5.3.1   Consumer demographics

Section 4.4, "Shelf Life Centered on the Product or on Its Interaction with the Consumer?" and the subsequent sections concluded that SSL should be centered on the probability of the consumer accepting or rejecting the stored product. A question to be asked is whether consumer demographics can have an influence on the probability of rejection. This question is now addressed with an example where the same set of yogurt samples was presented to a group of adults and to a group of children.

#### 5.3.1.1   Experimental data

To illustrate a model where the consumer demographics is taken as a covariate in predicting percent rejection, data from a yogurt storage test will be used. The product and storage conditions were described in Section 4.4. Basically a yogurt was stored at 42°C for 0, 4, 8, 12, 24, 36, and 48 hours.

Fifty adults aged between 18 and 30 years and 50 children aged between 10 and 12 years who consumed stirred yogurt at least once a week were recruited from the town of Nueve de Julio (Buenos Aires, Argentina). These consumers were presented with the seven yogurt samples monadically in random order. For each sample, subjects tasted the sample and answered the question: "Would you normally consume this product: Yes or No?" The tests were conducted in a sensory laboratory with individual booths with artificial daylight-type illumination, temperature control (between 22 and 24°C), and air circulation. The censored data are presented schematically in Table 5.3. The format of the data is similar to censored data presented in previous examples with the addition of a column corresponding to the covariate; that is the Group column in Table 5.3.

#### 5.3.1.2   Covariate model

In Section 4.7 the model to estimate the rejection function was presented. There it was said that usually rejection times are not normally distributed; instead, their distribution is often right skewed. Often, a loglinear model is chosen:

$$Y = \ln(T) = \mu + \sigma W, \tag{5.1}$$

where $W$ is the error term distribution.

*Table 5.3* Censored Data Obtained from Yogurt
Stored at 42°C

| Consumer[a] | tlow | thigh | cens | censcod | group |
|---|---|---|---|---|---|
| 2 | 4 | 24 | interval | 3 | adult |
| 3 | 48 | 48 | right | 0 | adult |
| 4 | 24 | 24 | left | 2 | adult |
| ... | ... | ... | ... | ... | ... |
| 48 | 4 | 4 | left | 2 | adult |
| 49 | 12 | 24 | interval | 3 | adult |
| 50 | 48 | 48 | right | 0 | adult |
| 1 | 8 | 48 | interval | 3 | child |
| 3 | 12 | 12 | left | 2 | child |
| 4 | 12 | 48 | interval | 3 | child |
| ... | ... | ... | ... | ... | ... |
| 48 | 48 | 48 | right | 0 | child |
| 49 | 48 | 48 | right | 0 | child |
| 50 | 48 | 48 | right | 0 | child |

*Note:* Data evaluated by a group of adults and a group of children. The full data can be downloaded from the editor's Web site: yogage.xls.

[a] Column headings: consumer, identification of consumers within each group; tlow, low time interval (hours); thigh, high time interval (hours); cens, type of censorship; censcod, censorship code; group: age group corresponding to each consumer.

To establish whether the consumers' age group influenced rejection times, the following loglinear regression model with inclusion of covariates was applied (Klein and Moeschberger 1997; Meeker and Escobar 1998). Its form is analogous to Equation 5.1:

$$Y = \ln(T) = \mu + \sigma W = \beta_0 + \beta_1 Z + \sigma W, \quad (5.2)$$

where $T$ is the storage time at which a consumer rejects a sample; $\beta_0$ and $\beta_1$ are the regression coefficients; $Z$ is the covariate indicating the group: adults ($Z = 0$) and children ($Z = 1$); $\sigma$ does not depend on the covariates; and $W$ is the error distribution. Regarding category covariates it is convenient to order the $Z$ values (0 and 1) in accordance with the alphabetical order of names (adults and children). This simplifies the interpretation of the R-output.

If the log-normal distribution is chosen for $T$, Equations 4.2 and 5.2 can be combined to obtain the following rejection function:

$$F(t) = \Phi\left(\frac{\ln(t) - \mu}{\sigma}\right) = \Phi\left(\frac{\ln(t) - (\beta_0 + \beta_1 Z)}{\sigma}\right). \qquad (5.3)$$

If the Weibull distribution is chosen for $T$, Equations 4.3 and 5.2 can be combined to obtain the following rejection function:

$$F(t) = 1 - \exp\left[-\exp\left(\frac{\ln(t) - \mu}{\sigma}\right)\right] = 1 - \exp\left[-\exp\left(\frac{\ln(t) - (\beta_0 + \beta_1 Z)}{\sigma}\right)\right]. \qquad (5.4)$$

### 5.3.1.3  Calculations using R

Statistical software used to analyze survival data with covariates is the same as that used to analyze interval-censored data with no covariates. As mentioned in Section 4.8, not all statistical software has the necessary procedures to deal with this type of censoring. R (http://www.r-project.org/, accessed May 26 2009) does have a procedure for interval-censoring calculations with covariates and is used to analyze the present type of data.

The first step in performing calculations is to enter the raw data in an Excel spreadsheet in a format as shown in Table 5.3. To obtain the data in this format the procedure can be as follows:

1. Obtain the raw data from consumers in the format as shown in Table 5.4—a separate table for each consumer group.
2. Process the raw data for each group using the R function *sslife* and store the results in, for example, resadul for the adult group.
3. Display the censored data by typing > resadul$censdata.
4. Select the censored data with the mouse and copy/paste the data to an Excel spreadsheet.
5. Using the *Data/Text in columns* utility, transform the pasted data to the corresponding columns.
6. Once both groups have been processed in this way, their censored data can be appended and the necessary columns shown in Table 5.3 can be created.

For the data to be compatible with the function written in R for SSL calculations, it should have the following characteristics:

1. The first column should indicate the consumer number. These numbers do not necessarily have to be consecutive nor start from 1, but all cells should be numbered and the column should have a text heading. In this example the heading is *consumer*, but it could be any other name. The data corresponding to 4 of the 50 adults was discarded

**Table 5.4** Acceptance/Rejection Data for Yogurt Stored
at 42°C for 6 of the 50 Adult Consumers

| Consumer | t0 | t4 | t8 | t12 | t24 | t36 | t48 |
|---|---|---|---|---|---|---|---|
| 1 | no | no | yes | yes | yes | yes | no |
| 2 | yes | yes | no | yes | no | no | no |
| 3 | yes | yes | no | yes | no | no | yes |
| ... | ... | ... | ... | ... | ... | ... | ... |
| 48 | yes | no | no | no | no | no | no |
| 49 | yes | yes | yes | yes | no | no | no |
| 50 | yes | yes | yes | yes | yes | yes | yes |

because they rejected the fresh sample; likewise for the data of 3 of the 50 children.

2. The second and third columns represent the storage times corresponding to the low and high times of the interval, respectively. The consumers do not have to have the same names as in Table 5.3, but the table must have a text heading.

3. The fourth and fifth columns represent the type of censoring that corresponds to each of the consumers, represented by a text (interval, right, or left) and a code (3, 0, and 2), respectively. They do not have to have the same names as in Table 5.3, but they must have a text heading.

4. The sixth column corresponds to the consumer groupings. It is recommended that these groupings are represented by alphanumerical characters and not numbers.

The data in the format shown in Table 5.3 should be saved as a tab-delimited text file. This format can be easily read by R. The instructions to analyze the data saved in the text file yogage.txt are the following:

1. Open R.
2. Go to the File Menu and change the directory, if necessary, to where the yogage.txt file was saved.
3. Go to the File Menu and open a New Script. This will bring up an empty window.
4. Introduce the text shown in Figure 5.4 in the New Script window. Once it has been introduced, save this window as *sslcat.R*.
5. What is sslcat.R? It is a function that will allow analyzing shelf-life data in a format such as shown in Table 5.3. The sslcat.R function has the following format and options:

```
sslcat <- function(daten,model="weibull",percent=
c(10,25,50))
```

```
sslcat <- function(daten,model="weibull",percent=c(10,25,50))
{
    library(survival)
    prop<-percent/100
    categ<-daten[,6]

    pp2<-survreg(Surv(daten[,2],daten[,3],daten[,5],type="interval")
~categ ,data=daten,dist=model)
    newcat<- factor(levels(categ))
    pp4<-predict(pp2,newdata=list(categ=newcat),type = "uquantile",
p = prop, se.fit = T)
    cat1<-levels(categ)
    catn<-nlevels(categ)
    nper<-length(percent)
    ci3<-array(dim=c(catn,nper,3))
    ci5<-list(dim=catn)
    parameter<-c("Intercept",cat1[2:catn],"sigma")
    catn1<-catn+1
    estimate<-c(1:catn1)
    lower<-c(1:catn1)
    upper<-c(1:catn1)
    znormal<-c(1:catn,NA)
    prob100<-c(1:catn,NA)

    for (i in 1:catn) {
        ci3[i,,] <- cbind(pp4$fit[i,],pp4$fit[i,] - 1.96 *
pp4$se.fit[i,],pp4$fit[i,] + 1.96 * pp4$se.fit[i,])
        if (model=="weibull"|model=="lognormal"|model=="loglogistic"|
model=="exponential") {ci3[i,,] <- exp(ci3[i,,])
        pp4$se.fit[i,]<-pp4$se.fit[i,]*ci3[i,,1]}
        ci2<-cbind(ci3[i,,],pp4$se.fit[i,])
        dimnames(ci2) <- list(Percent=percent,c("Estimate","Lower ci",
"Upper ci","Serror"))
        ci5[[i]]<-ci2
        estimate[i]<-pp2$coefficient[i]
        lower[i]<-pp2$coefficients[i] - 1.96 * sqrt(pp2$var[i,i])
        upper[i]<-pp2$coefficients[i] + 1.96 * sqrt(pp2$var[i,i])
        znormal[i]<-pp2$coefficients[i]/sqrt(pp2$var[i,i])
        prob100[i]<-pn<-(1-pnorm(abs(znormal[i])))*2*100
    }
    if (model=="exponential") {
        estimate[catn1]<-NA
        lower[catn1]<-NA
        upper[catn1]<-NA}
    else {
        estimate[catn1]<-pp2$scale
        lower[catn1]<-exp(log(pp2$scale)-1.96*sqrt(pp2$var[catn1,catn1]))
        upper[catn1]<-exp(log(pp2$scale)+1.96*sqrt(pp2$var[catn1,catn1]))
        }
    names(ci5)<-cat1

    chiprob<-(1-pchisq(-2*(pp2$loglik[1]-pp2$loglik[2]),catn-1))*100
    list(musig=data.frame(parameter,estimate,lower,upper,znormal,
prob100),loglike=-pp2$loglik,chiprob100=chiprob,slives=ci5)
}
```

**Figure 5.4** R-function for estimating sensory shelf life for censored data with categorical covariates.

- daten: censored data.
- model: parametric model of choice (weibull, exponential, gaussian, logistic, log-normal, or loglogistic).
- percent: percent rejection values for which I want estimated shelf lives. For example, if I want estimated shelf lives for percent rejections = 10%, 20%, 30%, 40%, and 50%, then percent = c(10,20,30,40,50). If I want a table with estimated shelf lives for a sequence of percent rejections from 10% to 50% at 1% increments, then percent = c (seq(10,50,by=1)).

6. Caveat: When text is copied from Word and pasted into R, the quotation marks are sometimes not read properly and have to be retyped in R.
7. Instructions in R are written in the R Console window after the > symbol; thus, to read the raw data the following instruction has to be written:

```
> age <-read.table("yogage.txt",header=TRUE)
```

- The < (*less than* symbol followed by a hyphen) is supposed to symbolize an arrow and is equivalent to an equal or assignment symbol.
- Alternatively you could read the data from a directory other than the working directory; for example:

```
> age<- read.table("C:\HOUGH\R_FILES\yogage.txt",
header=TRUE)
```

8. Go to the File Menu, Open Source R-code: sslcat.R.
9. For the yogage.txt data, use sslcat with the following options:

```
> resage <- sslcat(age,model="lognormal")
```

a. After executing the previous instruction by pressing Enter, the following message appears: Loading required package: splines.
b. The results of the SSL analysis are in a data structure called *resage*, which contains the model's parameters, the loglikelihood values, the significance level of the model, and the predicted shelf life values for each one of the covariates. Each one of these can be displayed separately by typing: resage$musig, resage$loglike, resage$chiprob100 and resage$slives; respectively.
c. Or they can all be displayed together by typing:

```
> resage
```

```
$musig
```

|   | parameter | estimate | lower | upper | znormal | prob100 |
|---|-----------|----------|-------|-------|---------|---------|
| 1 | Intercept | 2.9882021 | 2.7022322 | 3.274172 | 20.480745 | 0 |
| 2 | Child | 1.048621 | 0.6222476 | 1.474994 | 4.820416 | 0.00014326 |
| 3 | Sigma | 0.9071035 | 0.7294069 | 1.12809 | NA | NA |

```
$loglike
[1] 135.8571 124.2994

$chiprob100
[1] 0.0001525560

$slives

$slives$Adult
```

| percent | estimate | lower ci | upper ci | serror |
|---------|----------|----------|----------|--------|
| 10 | 6.207152 | 4.224986 | 9.119258 | 1.218268 |
| 25 | 10.765753 | 7.839564 | 14.784169 | 1.742222 |
| 50 | 19.849962 | 14.912983 | 26.421338 | 2.896169 |

```
$slives$Child
```

| percent | estimate | lower ci | upper ci | serror |
|---------|----------|----------|----------|--------|
| 10 | 17.71343 | 12.63867 | 24.82584 | 3.050704 |
| 25 | 30.72237 | 22.72288 | 41.53805 | 4.727778 |
| 50 | 56.6461 | 41.33757 | 77.62383 | 9.105321 |

13. Interpretation of the output:
    - $musig: The intercept corresponds to the $\beta_0$ value of Equation 5.2; as the log-normal distribution was chosen they also correspond to $\beta_0$ of Equation 5.3. The Child parameter corresponds to the $\beta_1$ value of Equation 5.2. To obtain $\mu$ for the Adult group, $Z = 0$ and to obtain $\mu$ for the Child group $Z = 1$. Remember that it was recommended above to order the $Z$ values (0 and 1) in accordance with the alphabetical order of their names (Adults and Children). The value of sigma is the $\sigma$ parameter of Equations 5.2 and 5.3. The $musig table presents the parameter estimates and their lower and upper 95% confidence limits. Also the normal curve z-values are presented together with their corresponding

significance levels. In this case both the Intercept and the Child parameter are significant.

- $loglike: These are the loglikelihood values corresponding to the reduced model without the covariate (higher loglike) and the full model including the covariate (lower loglike). The log-normal distribution was chosen because the loglikelihood value of the full model for this distribution was lower than for the other proposed distributions, although the loglikelihood for the loglogistic distribution was only slightly higher.

- $chiprob100: This is the significance level, expressed in %, of the experimental chi-square value obtained from the loglikelihood values: chi-square = 2 × (135.9 − 124.3) = 23.2 with 1 degree of freedom. The significance level is <1%; that is, the model with the covariate is highly significant and thus the percent rejection versus storage time curves for the Adult and Child groups are different.

- $slives: These tables display the estimated shelf lives corresponding to 10%, 25%, and 50% rejection probabilities for the Adult and Child groups. The table presents the estimates, the lower and upper 95% confidence intervals, and the standard error of the estimations.

With

$$\mu = \beta_0 + \beta_1 Z = 2.988 + 0 \times 1.0486 = 2.988$$

for the group of adults with

$$\mu = \beta_0 + \beta_1 Z = 2.988 + 1 \times 1.0486 = 4.037$$

for the group of children, and with $\sigma = 0.907$ for both groups, the rejection probability for the log-normal distribution given by Equation 5.3 can be calculated and graphed as shown in Figure 5.5. This graph clearly shows why the age effect was significant when calculating the parameters of the model with R. The censored data anticipated this result. The adult's group had 8 right-censored consumers while the children's group had 26. That is, more than half the children accepted the product with the maximum storage time which extended the SSL for this group in relation to adults. The difference between adults and children reinforces the concept expressed in Section 4.4 that food products do not have SSL of their own; rather these depend on the interaction of the product with the consumer.

Why did the children's group differ from the adults' group in their answers? In Section 2.7.3.1 the recruitment of children for preference tests

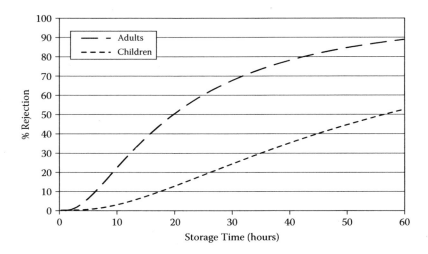

**Figure 5.5** Percent rejection versus storage time for the log-normal distribution corresponding to yogurt stored at 42°C for a group of adults and a group of children.

was discussed. One issue is that the sensory tests are relatively exciting experiences for children. They are taken to a special hall, and samples are presented with mysterious three-digit codes in odd sorts of containers like small plastic cups or sample holders. They are told they are going to receive a present when they finish their task. All this setup is far removed from their everyday eating experience, and the risk is that the result will be far removed from what they really feel about the product in a real setting. Another issue with the product is that stored yogurt has an acidic flavor. In a study using gelatins with different levels of citric acid, Liem and Mennella (2003) showed that more than one-third of the children, but virtually none of the adults, preferred the high levels of sour taste in gelatin. Thus, the longer SSL shown by the children's group could be due to their preferring more acidic yogurt than adults.

## 5.3.2   Product formulations

Different products, obviously, can have a different SSL. This will be reflected in acceptance/rejection patterns of consumers confronted with these different products. In some instances it can be of interest to estimate if the SSL of a current product is extended by a change in formulation or processing; it can also be of interest to contrast the SSL of the current product with the competitor product.

### 5.3.2.1 Experimental data

To illustrate a model where product is taken as a covariate in predicting percent rejection, data from a storage test of a type of dehydrated soup will be used. The products that were studied were *competitor, new,* and *traditional* formulations. All samples were stored at 23°C. The competitor product was stored for 0, 3, 7, 10, 11, 12, and 14 months; and the new and traditional products for 0, 3, 5, 8, 10, 12, and 14 months. A reversed storage design was used (see Section 3.3.6.2); thus, it was possible for each consumer to taste all the samples corresponding to a formulation in a single session in random order.

For each product type, 63 adult women who consumed the product category were recruited from the town of Nueve de Julio (Buenos Aires, Argentina); there were three product types, so this meant that 189 consumers were recruited in total. Each consumer was presented with the seven samples corresponding to different storage times monadically in random order. For each sample, subjects tasted the sample and answered the question: "Would you normally consume this product: Yes or No?" The tests were conducted in a sensory laboratory with individual booths with artificial daylight type illumination, temperature control (between 22 and 24°C), and air circulation. The censored data are presented schematically in Table 5.5. The format of the data is similar to censored data presented in previous examples with the addition of a column corresponding to the covariate; that is the *product* column in Table 5.5.

### 5.3.2.2 Calculations using R

In Section 5.3.1.2 the model to estimate the rejection function in the presence of covariates was presented, and the same model is used here. Suggestions to preparing the data to adjust to the format shown in Table 5.5 and the description of how the data should be presented are at the beginning of Section 5.3.1.3.

The data in the format shown in Table 5.5 should be saved as a tab-delimited text file. This format can be easily read by R. The instructions to analyze the data saved in the text file soup.txt are the following:

1. Open R.
2. Go to the File Menu and change the directory, if necessary, to where the soup.txt file was saved.
3. If the reader has already gone through Section 5.3.1, then the function sslcat.R has already been loaded. Otherwise, see Section 5.3.1.2.
4. To read the raw data, the following instruction has to be written:

```
> soup <-read.table("soup.txt",header=TRUE)
```

***Table 5.5*** Censored Data Obtained from Three Types
of Dehydrated Soup Stored at 23°C

| consumer[a] | tlow | thigh | cens | censcod | product |
|---|---|---|---|---|---|
| 4 | 14 | 14 | right | 0 | comp |
| 5 | 3 | 12 | interval | 3 | comp |
| 10 | 14 | 14 | right | 0 | comp |
| ... | ... | ... | ... | ... | comp |
| 63 | 5 | 14 | interval | 3 | comp |
| 64 | 3 | 5 | interval | 3 | new |
| 65 | 10 | 10 | left | 2 | new |
| 66 | 10 | 12 | interval | 3 | new |
| ... | ... | ... | ... | ... | new |
| 125 | 14 | 14 | right | 0 | new |
| 126 | 3 | 3 | left | 2 | trad |
| 127 | 8 | 8 | left | 2 | trad |
| 129 | 14 | 14 | right | 0 | trad |
| ... | ... | ... | ... | ... | trad |
| 189 | 8 | 14 | interval | 3 | trad |

*Note:* Data evaluated by a group of adults. The full data can be
downloaded from the editor's Web site: soup.xls.

[a] Column headings: consumer, identification of consumers within
each product; tlow: low time interval (months); thigh: high time
interval (months); cens: type of censorship; censcod: censorship
code; product: product type, comp = competitor, new = new formulation, trad = traditional formulation.

- Alternatively you could read the data from a directory other than
  the working directory, for example:

```
> soup <- read.table("C:\HOUGH\R_FILES\soup.
txt",header=TRUE)
```

5. Go to the File Menu, Open Source R-code: sslcat.R.
6. For the soup.txt data, use sslcat with the following options:

```
> ressoup <- sslcat(soup,model="gaussian")
```

7. After executing the previous instruction by pressing Enter, the following message appears: Loading required package: splines.
8. The results of the SSL analysis are in a data structure called *ressoup*,
   which contains the model's parameters, the loglikelihood values, the significance level of the model, and the predicted shelf
   life values for each one of the covariates. Each one of these can be

displayed separately by typing: ressoup$musig, ressoup$loglike, ressoup$chiprob100, and ressoup$slives,; respectively.

9. Or, they can all be displayed together by typing:

```
> ressoup
```

```
$musig
```

| | parameter | estimate | lower | upper | znormal | prob100 |
|---|---|---|---|---|---|---|
| 1 | intercept | 15.444757 | 12.3734577 | 18.516056 | 9.856325 | 0 |
| 2 | new | 4.379743 | 0.08739895 | 8.672088 | 1.999909 | 4.551012 |
| 3 | trad | 2.488045 | −1.46491584 | 6.441005 | 1.233649 | 21.733356 |
| 4 | sigma | 7.818838 | 5.78400966 | 10.569525 | NA | NA |

```
$loglike
[1] 125.0271 122.8824
```

```
$chiprob100
[1] 11.71040
```

```
$slives
```

```
$slives$comp
```

| percent | estimate | lower ci | upper ci | serror |
|---|---|---|---|---|
| 10 | 5.424512 | 2.045319 | 8.803705 | 1.724078 |
| 25 | 10.17103 | 7.306832 | 13.035229 | 1.461326 |
| 50 | 15.444757 | 12.373458 | 18.516056 | 1.566989 |

```
$slives$new
```

| percent | estimate | lower ci | upper ci | serror |
|---|---|---|---|---|
| 10 | 9.804256 | 6.622815 | 12.9857 | 1.623184 |
| 25 | 14.550774 | 11.429207 | 17.67234 | 1.592636 |
| 50 | 19.8245 | 16.065968 | 23.58303 | 1.917619 |

```
$slives$trad
```

| percent | estimate | lower ci | upper ci | serror |
|---|---|---|---|---|
| 10 | 7.912557 | 4.823336 | 11.00178 | 1.576133 |
| 25 | 12.659075 | 9.931684 | 15.38647 | 1.391526 |
| 50 | 17.932801 | 14.786062 | 21.07954 | 1.605479 |

14. Interpretation of the output:

- $musig: The intercept corresponds to the $\beta_0$ value of Equation 5.2. The *new* and *trad* parameters correspond to the $\beta_1$ value of Equation 5.2. To obtain $\mu$ for the competitor product, $Z = 0$. To obtain $\mu$ for the new product group, $\beta_1 = 4.38$ and $Z = 1$; to obtain $\mu$ for the traditional product group, $\beta_1 = 2.488$ and $Z = 1$. The value of sigma is the $\sigma$ parameter of Equations 5.2 and 5.3. The $musig table presents the parameter estimates and their lower and upper 95% confidence limits. Also, the normal curve z-values are presented together with their corresponding significance levels. In this case the Intercept and the *new* product are significant; that is there is a tendency for the new product to have a different shelf life in comparison to the competitor product.
- $loglike: These are the loglikelihood values corresponding to the reduced model without the covariate (higher loglike) and the full model including the covariate (lower loglike). The Gaussian distribution was chosen because the loglikelihood value of the full model for this distribution was lower than for the other proposed distributions, although the loglikelihood for the Weibull distribution was only slightly higher.
- $chiprob100: This is the significance level, expressed in %, of the experimental chi-square value obtained from the loglikelihood values: chi-square = $2 \times (125.02 - 122.88) = 4.28$ with 2 degrees of freedom. The significance level is 11.7%; that is, the model with the covariate is not strictly significant. In the previous paragraph it was shown that the new product tended to have a rejection curve different from the competitor product. From the chi-square test on the loglikelihoods this can only be taken as a tendency. If the new product is not too costly to produce, with this data it could be adopted.
- $slives: These tables display the estimated shelf lives corresponding to 10%, 25%, and 50% rejection probabilities for each one of the products. The table presents the estimates, the lower and upper 95% confidence intervals, and the standard error of the estimations.

With

$$\mu = \beta_0 = 15.445$$

for the competitor product; with

$$\mu = \beta_0 + \beta_1 Z = 15.445 + 1 \times 4.3797 = 19.825$$

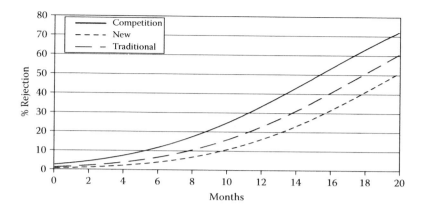

*Figure 5.6* Percent rejection versus storage time for the normal distribution corresponding to three types of dehydrated soup stored at 23°C evaluated by adult consumers.

for the new product; with

$$\mu = \beta_0 + \beta_1 Z = 15.445 + 1 \times 2.488 = 17.933$$

for the traditional product; and with $\sigma = 7.819$ for the three products, the rejection probability for the normal distribution can be calculated by:

$$F(t) = \Phi\left(\frac{t-\mu}{\sigma}\right) = \Phi\left(\frac{t-(\beta_0 + \beta_1 Z)}{\sigma}\right) \quad (5.5)$$

The normal distribution in Excel is expressed as:

=DISTR.NORM(C4,$B$12,$B$13,TRUE)*100

where C4: time, B12: $\mu$, and B13: $\sigma$. The resulting graph for the three products is in Figure 5.6.

The general chi-square test based on the loglikelihood is valid for the overall rejection function. However, it could be of interest to test for differences of predicted SSL at specific percent rejections. For example, is the estimated SSL for 25% rejection for the new formulation significantly higher than for the competitor product? For these comparisons the following equation can be used (Meeker and Escobar 1998):

$$CI = \text{shelf life}_{new} - \text{shelf life}_{competitor} \pm Z_{(1-\alpha/2)} se_{difference} \quad (5.6)$$

where $Z_{(1-\alpha/2)}$ is the $(1-\alpha/2)$-quantile of the standard normal distribution and $se_{\text{difference}}$ is the standard error of the shelf-life difference, calculated by the following equation:

$$se_{\text{difference}} = \sqrt{\frac{1}{2}\frac{n_1+n_2}{n_1 n_2}}\sqrt{se_{\text{new}}^2 n_1 + se_{\text{competitor}}^2 n_2} \tag{5.7}$$

where $n_1$ and $n_2$ are the numbers of consumers who tasted the new formulation and competitor soups, respectively, and $se_{\text{new}}$ and $se_{\text{competitor}}$ are the standard errors of the shelf lives of the new and competitor soups, respectively.

Of the 63 consumers who evaluated the new product, 18 rejected the fresh sample; and of the 63 consumers who evaluated the competitor product, 23 rejected the fresh sample. This relatively high number of consumers rejecting the fresh sample could be due to the type of product, which can be eaten alone but is usually used as a broth in addition to other meals. Also, the soups were evaluated in the summer, and this could have generated a larger number of rejections for the fresh sample than expected. Thus $n_1 = 45$ and $n_2 = 40$. From the results of the R calculations shown above shelf-life$_{\text{new}}$ = 14.55, shelf-life$_{\text{competitor}}$ = 10.17, $se_{\text{new}}$ = 1.593 and $se_{\text{competitor}}$ = 1.461. Introducing these values in Equations 5.6 and 5.7:

$$\text{CI} = 14.55 - 10.17 \pm 1.96 \times 2.17 = 4.38 \pm 4.25$$

The confidence interval does not include 0; thus, it can be concluded that the SSL for 25% rejection is significantly higher ($P < 5\%$) for the new formulation in relation to the competitor's product.

Another possibility in the use of Equations 5.5 and 5.6 is to determine the significance level at which the SSL for two products differ. For example, we can compare the SSL for 25% rejection of the traditional formulation with the competitor's product. Thus $n_1 = 49$ and $n_2 = 40$. From the results of the R calculations shown above, shelf-life$_{\text{traditional}}$ = 12.66, shelf-life$_{\text{competitor}}$ = 10.17, $se_{\text{traditional}}$ = 1.39 and $se_{\text{competitor}}$ = 1.461. Introducing these values in Equation 5.6 and 5.7:

$$\text{CI} = 12.66 - 10.17 \pm Z_{(1-\alpha/2)} \times 2.023 = 2.49 \pm Z_{(1-\alpha/2)} \times 2.023$$

For the products to differ the confidence interval has to include 0. The $\alpha$ value for which $Z_{(1-\alpha/2)} \times 2.023 = 2.49$ is 0.218; thus the significance level at which the SSLs of the traditional versus the competitor products differ is 21.8%.

## 5.3.3 Quantitative covariates and number of covariates

In the examples presented in Sections 5.3.1 and 5.3.2, the covariates were categorical. In the first case there were two consumer age groups: adults and children. In the second case there were three products: competitor, new, and traditional. The covariate can also be a numerical continuous variable as shown in the following example.

### 5.3.3.1 Experimental data

In Section 4.10 an example was presented on the study of the SSL of fat-free stirred strawberry yogurt stored at a temperature of 10°C. Storage times at 10°C were 0, 14, 28, 42, 56, 70, and 84 days. Eighty people who consumed fat-free stirred yogurt at least once a week were recruited from the city of Nueve de Julio, Buenos Aires, Argentina. Each consumer received the seven yogurt samples (corresponding to each storage time at 10°C) monadically in random order. For each sample subjects had to answer the question "Would you normally consume this product: Yes or No?" Additionally, consumers were asked to score the acceptability of each sample using a numbered scale from 1 (*dislike very much*) to 10 (*like very much*). The data for six of these consumers is in Table 5.6. An extra column has been added to include each consumer's average acceptability of the seven samples corresponding to the seven storage times. As mentioned in

*Table 5.6* Censored and Acceptability Data for Six Consumers Who Evaluated Fat-Free Stirred Yogurt Stored at 10°C

| consumer[a] | tlow | thigh | cens | censcod | accept |
|---|---|---|---|---|---|
| 1 | 14 | 56 | interval | 3 | 5.14 |
| 3 | 14 | 70 | interval | 3 | 6.86 |
| 4 | 14 | 70 | interval | 3 | 6.43 |
| 14 | 28 | 42 | interval | 3 | 4.57 |
| 17 | 56 | 70 | interval | 3 | 6.57 |
| ... | ... | ... | ... | ... | ... |
| 78 | 28 | 42 | interval | 3 | 6.14 |
| 79 | 42 | 56 | interval | 3 | 5.71 |
| 80 | 28 | 42 | interval | 3 | 5.86 |

*Note:* The full data can be downloaded from the editor's Web site: acceptability.xls.

[a] Column headings: consumer, identification of consumers; tlow, low time interval (days); thigh, high time interval (days); cens, type of censorship; censcod, censorship code; accept, average acceptability score (1, *dislike very much*; 10, *like very much*) for the seven samples corresponding to seven storage times.

Section 4.8, to process the data with R, the data have to be stored in tab-delimited text file; we store this data in acceptability.txt.

### 5.3.3.2   Calculations using R
In Section 5.3.1.2 the model to estimate the rejection function in the presence of covariates was presented:

$$Y = \ln(T) = \mu + \sigma W = \beta_0 + \beta_1 Z + \sigma W \qquad (5.8)$$

where $T$ is the storage time at which a consumer rejects a sample; $\beta_0$ and $\beta_1$ are the regression coefficients; $Z$ is the covariate indicating the acceptability; $\sigma$ does not depend on the covariates; and $W$ is the error distribution.

If, for example, the normal distribution is chosen for $T$, the following rejection function can be obtained:

$$F(t) = \Phi\left(\frac{t - \mu}{\sigma}\right) = \Phi\left(\frac{t - (\beta_0 + \beta_1 Z)}{\sigma}\right) \qquad (5.9)$$

For the data to be compatible with the function written in R for SSL calculations, it should have the following characteristics:

1. The first column should indicate the consumer number.
2. The second and third columns represent the storage times corresponding to the low and high times of the interval, respectively.
3. The fourth and fifth columns represent the type of censoring that corresponds to each of the consumers, represented by a text (Interval, Right, or Left) and a code (3, 0, and 2), respectively.
4. The sixth column corresponds to the quantitative numerical covariate, in this case, the average consumer acceptability.

The instructions to analyze the data saved in the text file acceptability.txt are the following:

1. Open R.
2. Go to the File Menu and change the directory, if necessary, to where the acceptability.txt file was saved.
3. Go to the File Menu and open a New Script. This will bring up an empty window.
4. Introduce the text shown in Figure 5.7 in the New Script window. Once it has been introduced, save this window as sslcov.R.

```
sslcov <- function(daten,predict=1,model="weibull",percent=c(10,25,50))
{
    library(survival)
    prop<-percent/100
    kel<-daten[,6]
    pp2<-survreg(Surv(daten[,2],daten[,3],daten[,5],type="interval")
~kel ,data=daten,dist=model)
    newkel<-predict
    pp4<-predict(pp2,newdata=list(kel=newkel),type = "uquantile",
p = prop, se.fit = T)

    ci3 <- cbind(pp4$fit,pp4$fit - 1.96 * pp4$se.fit,pp4$fit + 1.96 *
pp4$se.fit)
    if
(model=="weibull"|model=="lognormal"|model=="loglogistic"|model==
"exponential") {
    ci3 <- exp(ci3)
    pp4$se.fit<-pp4$se.fit*ci3[,1]}
    ci2<-cbind(ci3,pp4$se.fit)
    beta0 <- c(pp2$coefficients[1],pp2$coefficients[1] - 1.96 *
sqrt(pp2$var[1,1]),pp2$coefficients[1] + 1.96 * sqrt(pp2$var[1,1]))
    beta1<-c(pp2$coefficients[2],pp2$coefficients[2] - 1.96 *
sqrt(pp2$var[2,2]),pp2$coefficients[2] + 1.96 * sqrt(pp2$var[2,2]))
    if (model==»exponential») {
    sigma<-c(NA,NA,NA)}
    else {
    sigma<-c(pp2$scale,exp(log(pp2$scale)-1.96*sqrt(pp2$var[3,3])),
exp(log(pp2$scale)+1.96*sqrt(pp2$var[3,3])))
    }
    dimnames(ci2) <- list(Percent=percent,c("Estimate","Lower ci",
"Upper ci","Serror"))

    value<-c(«estimate»,»lower»,»upper»)
    chiprob<-(1-pchisq(-2*(pp2$loglik[1]-pp2$loglik[2]),1))*100
    list(musig=data.frame(value,beta0,sigma,beta1),loglike=
-pp2$loglik,chiprob100=chiprob,predict.for=predict,slives=ci2)
}
```

*Figure 5.7* R-function for estimating sensory shelf life for censored data with quantitative covariates.

5. What is sslcov.R? It is a function that will allow analyzing shelf-life data in a format such as shown in Table 5.6. The sslcov.R function has the following format and options:

```
sslcov <- function(daten,predict=1,model="weibull",
percent=c(10,25,50))
```

- daten: censored data.
- predict: numerical value of the covariate for which I want estimated shelf lives.
- model: parametric model of choice (weibull, exponential, gaussian, logistic, log-normal, or loglogistic).

- percent: percent rejection values for which I want estimated shelf lives. For example, if I want estimated shelf lives for percent rejections = 10%, 20%, 30%, 40%, and 50%, then percent=c(10,20,30,40,50). If I want a table with estimated shelf lives for a sequence of percent rejections from 10% to 50% at 1% increments, then percent=c(seq(10,50,by=1)).

6. Instructions in R are written in the R Console window after the > symbol; thus, to read the raw data the following instruction has to be written:

```
> accept <-read.table("acceptability.txt",
header=TRUE)
```

- Alternatively you could read the data from a directory other than the working directory, for example:

```
> accept<- read.table("C:\HOUGH\R_FILES\
acceptability.txt",header=TRUE)
```

7. Go to the File Menu, Open Source R-code: sslcov.R.
8. For the acceptability.txt data, use sslcov with the following options:

```
> resaccept <- sslcov(accept,predict=5.9,model=
"gaussian")
```

9. After executing the previous instruction by pressing Enter, the following message appears: Loading required package: splines.
10. The results of the SSL analysis are in a data structure called *resaccept*, which contains the model's parameters, the loglikelihood values, the significance level of the model, and the predicted shelf life values for the level of the covariate introduced in the sslcov function; in this case acceptability = 5.9. Each one of these can be displayed separately by typing: resaccept$musig, resaccept$loglike, resaccept$chiprob100, and resaccept$slives, respectively.
11. Or they can all be displayed together by typing:

```
> resaccept
```

```
$musig
```

|   | value | beta0 | sigma | beta1 |
|---|-------|-------|-------|-------|
| 1 | estimate | −25.97468 | 11.618133 | 12.316598 |
| 2 | lower | −46.15384 | 9.370763 | 9.094676 |
| 3 | upper | −5.79551 | 14.404485 | 15.53852 |

```
$loglike

[1]  84.11572  62.46575

$chiprob100

[1]  4.695955e-09

$predict.for

[1]  5.9

$slives
```

| percent | estimate | lower ci | upper ci | serror |
|---:|---:|---:|---:|---:|
| 10 | 31.80401 | 26.79179 | 36.81624 | 2.557256 |
| 25 | 38.85694 | 34.8068 | 42.90708 | 2.066398 |
| 50 | 46.69325 | 43.21352 | 50.17298 | 1.775372 |

12. Interpretation of the output:
   - $musig: The beta0 corresponds to the $\beta_0$ value of Equation 5.8; as the normal (or Gaussian) distribution was chosen, they also correspond to $\beta_0$ of Equation 5.9. The beta1 parameter corresponds to the $\beta_1$ value of Equations 5.8 and 5.9. The value of sigma is the $\sigma$ parameter of Equations 5.8 and 5.9. The $musig table presents the parameter estimates and their lower and upper 95% confidence limits.
   - $loglike: These are the loglikelihood values corresponding to the reduced model without the covariate (higher loglike) and the full model including the covariate (lower loglike). The normal distribution was chosen because the loglikelihood value of the full model for this distribution was lower than for the other proposed distributions.
   - $chiprob100: This is the significance level, expressed in %, of the experimental chi-square value obtained from the loglikelihood values: chi-square = 2 × (84.1 − 62.5) = 43.2 with 1 degree of freedom. The significance level is <1%; that is, the model with the covariate is highly significant, and thus the percent rejection versus storage time curves depend on the average acceptability.
   - $slives: These tables display the estimated shelf lives corresponding to 10%, 25%, and 50% rejection probabilities for the chosen value of the covariate used in the *predict* option of the function sslcov; in this case predict = 5.9. The table presents the estimates, the lower and upper 95% confidence intervals, and the standard error of the estimations.

The acceptability average per consumer shown in the last column of Table 5.6 has the following distribution characteristics: mean = 6.0, median = 5.9, lower quartile = 5.3, and upper quartile = 6.9. To calculate the $\mu$ value for the normal distribution corresponding to the median:

$$\mu = \beta_0 + \beta_1 Z = -25.97 + 12.317 \times 5.9 = 46.7$$

With this value of $\mu = 46.7$ and $\sigma = 11.62$, equation 5.9 can be used to graph percent rejection versus storage time for consumers who presented an average acceptability = 5.9. This graph and the curves corresponding to the lower- and upper-quartile average acceptabilities are in Figure 5.8. The results obtained from this data are as would be expected. Consumers who had low average-acceptability values were more likely to be those that rejected samples at lower storage times than those consumers who had higher average-acceptability values. Table 5.7 shows the acceptability and accept/reject data for two of the consumers involved in the study, identified as id = 14 and 17. Consumer 14, who was interval-censored between 28 and 42 days storage, had an average acceptability of 4.57; while consumer 17, who was interval-censored between 56 and 70 days storage, had an average acceptability of 6.57. What is clear from this example is that for this product there was a close relationship between acceptance/rejection of a sample and the corresponding acceptability score.

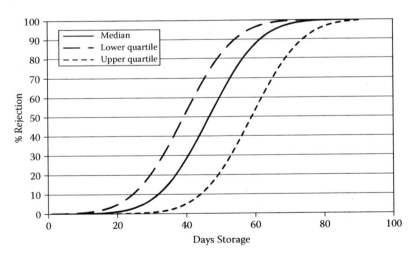

*Figure 5.8* Percent rejection versus storage time for fat-free strawberry yogurt stored at 10°C for consumers with different average acceptabilities on a 1–10 acceptability scale: median = 5.9, lower quartile = 5.3, and upper quartile = 6.9.

**Table 5.7** Acceptance/Rejection and Acceptability Data for Two Consumers Who Evaluated Fat-Free Stirred Yogurt Stored at 10°C

| Storage time (days) | Consumer 14 | | Consumer 17 | |
|---|---|---|---|---|
| | Accept | Acceptability score | Accept | Acceptability score |
| 0 | Yes | 7 | Yes | 6 |
| 14 | Yes | 8 | Yes | 8 |
| 28 | Yes | 8 | Yes | 7 |
| 42 | No | 6 | Yes | 9 |
| 56 | No | 1 | Yes | 9 |
| 70 | No | 1 | No | 4 |
| 84 | No | 1 | No | 3 |

## 5.3.4   Number of covariates

The previous examples have included a single covariate, categorical or numerical. There are situations where there is more than one covariate. In Chapter 8 an example will be presented where there were three covariates in a study to determine the optimum internal cooking temperature for beef. The sslcat and sslcov R-functions shown in Figures 5.4 and 5.7, respectively, are designed to be used with a single covariate. The interested R user can introduce more than one covariate in the *survreg* R-function.

# References

Araneda, M., G. Hough, and E. Wittig de Penna. 2008. Current-status survival analysis methodology applied to estimating sensory shelf life of ready-to-eat lettuce (*Lactuta sativa*). *Journal of Sensory Studies* 23: 162–170.

Curia, A., M. Aguerrido, K. Langohr, and G. Hough. 2005. Survival analysis applied to sensory shelf life of yogurts. I: Argentine formulations. *Journal of Food Science* 70: S442–445.

Gámbaro, A., A. Giménez, P. Varela, L. Garitta, and G. Hough. 2004. Sensory shelf-life estimation of "alfajor" by survival analysis. *Journal of Sensory Studies* 19: 500–509.

Hough, G., M.L. Calle, C. Serrat, and A. Curia. 2007. Number of consumers necessary for shelf-life estimations based on survival analysis statistics. *Food Quality and Preference* 18: 771–775.

Hough, G., K. Langohr, G. Gómez, and A. Curia. 2003. Survival analysis applied to sensory shelf life of foods. *Journal of Food Science* 68: 359–362.

Hough, G., L. Garitta, and G. Gómez. 2006a. Sensory shelf life predictions by survival analysis accelerated storage models. *Food Quality and Preference* 17: 468–473.

Hough, G., I. Wakeling, A. Mucci, E. Chambers IV, I. Méndez Gallardo, and L. Rangel Alves. 2006b. Number of consumers necessary for sensory acceptability tests. *Food Quality and Preference* 17: 522–526.

Klein, J.P., and M.L. Moeschberger. 1997. *Survival analysis, techniques for censored and truncated data.* New York: Springer-Verlag Inc.

Liem, D.G., and J.A. Mennella. 2003. Heightened sour preferences during childhood. *Chemical Senses* 28: 173–180.

Meeker, W.Q., and L.A. Escobar. 1998. *Statistical methods for reliability data.* New York: John Wiley & Sons.

Salvador, A., S.M. Fiszman, A. Curia, and G. Hough. 2005. Survival analysis applied to sensory shelf life of yogurts. II: Spanish formulations. *Journal of Food Science* 70: S446–449.

Santa Cruz, M.J., L. Garitta, and G. Hough. 2001. Sensory descriptive analysis of yerba mate (*Ilex paraguariensis* Saint Hilaire), a South American beverage. *Food Science and Technology International* 7:1–7.

Shiboski, C. 1998. Generalized additive models for current status data. *Lifetime Data Analysis* 4: 29–50.

# chapter 6

# Cut-off point (COP) methodology

## 6.1    When is the survival statistics methodology difficult to apply?

The survival statistics methodology, based on consumers' acceptance or rejection of stored samples, has been described in detail in Chapters 4 and 5. As stated in Section 4.9, there are two main advantages to survival analysis methodology. One is that experimental sensory work is relatively simple. A group of consumers is asked to evaluate the stored samples by accepting or rejecting them; there is no need for sophisticated trained sensory panels or complicated consumer instructions. The second advantage is that the information obtained from consumers by this method is directly related to their everyday eating experience. When consumers are confronted with a food product, they either accept or reject it, just as is asked of them using this methodology.

Because of the above stated advantages the survival analysis methodology is recommended for sensory shelf life (SSL) determinations. But there are situations where it is difficult to apply. The foremost difficulty arises when there are several covariables, each with several levels. For example, in a study on the SSL of a vegetable oil, the following covariables were of interest:

1. Three storage temperatures (this particular covariable will be treated in-depth in Chapter 7)
2. Two packaging materials: regular PET and PET with an ultraviolet radiation filter in its formulation
3. Two illumination modes: dark and illumination 12 hours per day with fluorescent lamps to simulate exposure in a supermarket

With these three covariables, the total number of treatments was 3 temperatures × 2 packaging materials × 2 illumination modes = 12 treatments. For each one of these treatments there were six storage times.

In Section 5.1 the number of necessary consumers for the survival analysis was discussed; considering accepted statistical parameters a number of 120 consumers was recommended (Hough et al. 2007). Thus, for the 12 treatments a total of 12 × 120 = 1440 consumers would have to be recruited. This is not a reasonable proposal.

Could the same 120 consumers evaluate the 12 treatments in different sessions? This may be convenient, but not advisable. After the first two or three treatments the consumers would know that among the samples corresponding to the six storage times of each treatment are some to be rejected. They also acquire a degree of training in detecting the off-flavors developed during storage.

There are other cases in which the number of covariables and/or levels is not high, yet there is difficulty in recruiting consumers. Curia and Hough (2009) presented a study on the SSL of a fluid human milk replacement formula. They performed the study at three temperatures. This would require, as discussed in Section 5.1 and in Hough et al. (2007), 120 consumers for each temperature condition, which is a total of 360 consumers. These consumers are not the babies who consume the milk, but rather the mothers or caretakers who administer it. As discussed by Curia and Hough (2009) it was difficult to find mothers who were giving their babies this type of milk and were willing to perform the test. Thus a bare minimum of consumers were recruited. For survival analysis, suppose that a bare minimum of 50 consumers per storage temperature were recruited. This would mean a total of 150 mothers who fed their babies this sort of human milk replacement formula. In our town, which has a total of 40,000 inhabitants, this would be impossible. Considering largely populated cities the recruitment could be done, but it would be expensive and most probably beyond the budget of a SSL test. So this is a case where the numbers of covariables and levels were small, yet the difficulty in consumer recruitment is a barrier to the application of survival analysis methodology.

## 6.2   Basics of the COP methodology

To illustrate the ideas of the COP methodology, suppose we are to measure the SSL of sunflower oil. Samples are stored at 45°C for 90 days, and every 8 to 10 days a trained sensory panel measures oxidized flavor versus a control sample stored at 4°C. Figure 6.1 shows how oxidized flavor changes during storage time. To be able to establish the SSL, some decision has to be made regarding the maximum level of oxidized flavor that will be tolerated by consumers. If the maximum level is determined as 2 on the 0 to 10 sensory scale, then the estimated shelf life is 25 days; if the maximum level is 4, then the estimated shelf life is 70 days. The key issue is how to establish the maximum level, which we will call the COP. Different approaches presented in the literature to estimate the COP will be discussed, and a methodology to calculate a COP will be presented. Once the COP has been determined, then it can be applied to different

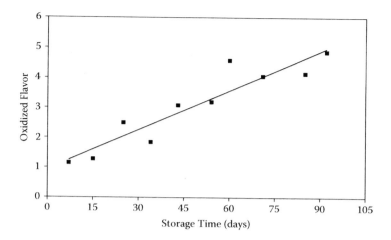

**Figure 6.1** Oxidized flavor for sunflower oil stored at 45°C with 12 hours illumination per day.

covariates and their different levels without the necessity of recruiting consumers, as would be the case for the survival analysis methodology.

## 6.3   Approaches in establishing a COP

One of the most frequently used approaches in establishing a COP is for the researcher to decide on an arbitrary value. A literature search in peer-reviewed international journals revealed that between 2006 and 2008 there were at least 25 papers dealing with SSL where an arbitrary COP was used. Following are three examples:

- Patsias et al. (2006): A panel of seven trained assessors measured acceptability of chicken breasts as a composite of odor and taste using a descriptive scale ranging from 1 to 9, where 1 = *extreme foreign flavor* or *dislike intensely* and 9 = *no foreign flavor* or *like extremely*. A score of 6 was taken as the lower limit of acceptability. The rule that trained assessors should not measure acceptability (see Table 2.1 in Chapter 2) was not complied with in this work. Also, no reason was given as to why the limit of 6 was chosen.
- Nunes et al. (2006): Shriveling of papaya fruit was determined by an undetermined number of assessors using a 1 to 5 visual rating scale where 1 = *field-fresh, no signs of shriveling*, 2 = *minor signs of shriveling*, 3 = *shriveling evident but not serious*, 4 = *moderate shriveling*, and 5 = *extremely wilted and dry*. Without any apparent justification, a

shriveling rating of 3 was considered to be the limit of acceptability for sale.

- Gómez-Guillén et al. (2007): Odor, melanosis, color of the head (cephalothorax), the cephalotorax–tail junction and the tail–parapods junction, and the general acceptability of Norway lobster were measured by a panel of eight trained assessors. They used scales ranging from 5 (*very fresh*) to 0 (*very spoiled*), fixing the rejection point at ≤2 based on previous trials. There was no description of previous trials.

Muñoz et el. (1992) presented case studies where a value of 6.5 on a 1 to 9 acceptability scale was taken as the minimum acceptability to establish a COP for quality control specifications. This value has been used by a number of authors as a means of determining a COP.

Ramírez et al. (2001) performed a SSL study on sunflower oil. They argued that a consumer panel would be the most appropriate one to determine when a food product reaches the end of its shelf life. To repeatedly assemble consumer panels for the multiple measurements needed during shelf-life studies would not be practical in many cases. A trained sensory panel is a lot simpler to assemble but yields analytical answers such as degree of oxidized flavor. How high does the oxidized flavor need to be for the sensory acceptability of the product to decrease? The answer to this question can be obtained by correlating data obtained from a consumer panel with data obtained from a trained panel. Ramírez et al. (2001) obtained the COP using the following procedure:

1. Estimating consumer's sensory failure: Average acceptability of fresh sample was based on LSD = 6.5 – 0.7 = 5.8, where LSD was Fisher's 5% least significance level obtained from the analysis of variance of the consumer data, considering consumers as the blocking factor and samples with different degrees of oxidized flavor as the treatment factor.
2. Correlating consumer acceptability for samples with different degree of oxidized flavor versus oxidized flavor measured on the same samples by a trained sensory panel as shown in Figure 6.2. Entering with an acceptability value of 5.8, the oxidation COP was estimated as 3.6 on the 0 to 10 scale.

Other authors have based their COPs on consumer studies as done by Ramírez et al. (2001). Hough et al. (2002) determined the COPs of sensory defects in powdered milk; Garitta et al. (2004) determined the COPs of burnt and plastic flavor defects in dulce de leche; and Curia and Hough (2009) measured the COP of dark color in a human milk replacement formula. The approach followed by these authors shall be presented in the following sections.

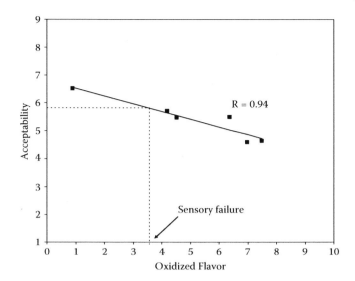

**Figure 6.2** Average acceptability of 51 consumers versus oxidized flavor scored by a trained panel. (Reprinted with permission from: Ramírez, G., G. Hough, and A. Contarini. 2001. Influence of temperature and light exposure on sensory shelf life of a commercial sunflower oil. *Journal of Food Quality* 24: 195–204.)

## 6.4  Methodology to measure the COP

The following stages are necessary to measure the COP of a sensory defect: (a) determination of the critical descriptors, (b) preparation of a series of samples with an increasing level of the sensory defect, (c) determination of the intensity level of the samples by a trained sensory panel, (d) determination of the acceptability level of the same samples by a consumer panel, and (e) calculation of the COP.

### 6.4.1  Critical descriptors

The issue of critical descriptors was discussed in detail in Section 3.3.4. The case study to be followed in describing COP methodology will be on the SSL of dulce de leche (DL) (Garitta et al. 2004). DL is a typical Argentine dairy product prepared from milk that is concentrated by evaporation with added sucrose and glucose (Moro and Hough 1985). For this product the critical descriptors were obtained by storing DL at 45°C during 30 days. Every 4–5 days a trained sensory panel compared the stored product versus a fresh control stored at 4°C. The descriptors that changed most during accelerated storage were plastic flavor, burnt flavor, dark color, and spreadability. Of these, plastic flavor was the first that reached its COP; thus it will be considered the critical descriptor.

## 6.4.2  Preparation of samples with increasing levels of sensory defects

To prepare samples with an increasing level of a sensory defect, there are basically two approaches:

1. Store samples, usually at an accelerated condition, at different times to develop the sensory defect.
2. Prepare samples by addition of a specific compound that produces the defect or by dilution of a sample where the defect is very pronounced.

For the case of DL the plastic flavor defect was achieved by heating 250 g of DL in its polystyrene pot for 24 hours at 80°C. Every 6 hours the DL was stirred in the pot for approximately 2 minutes. This stock reference was diluted in untreated DL to obtain different intensities of the defect; the following percent stock references were used to prepare nine samples with an increasing level of plastic flavor: 0, 6, 9, 13, 20, 30, 44, 67, and 100%.

Other examples of sensory defect preparations are as follows:

- Dark color in reconstituted milk powder (Hough et al. 2002): 100 ml of 2% coloring (SICNA, Milan, Italy caramel coloring) solution completed to 1 liter with reconstituted milk powder. This was the stock solution that was diluted to prepare different levels of the defect.
- Samples of sunflower oil with different levels of oxidized flavor were prepared by storing bottled oil at 60°C during 0, 15, 30, 45, 60, and 75 days (Ramírez et al. 2001).

## 6.4.3  Determination of intensity levels of samples by a trained sensory panel

A nine-member trained sensory panel was calibrated in the plastic flavor descriptor. Assessors received samples with 0, 6, 20, and 100% stock reference (see previous section). After discussion these samples were given scores of 0, 10, 50, and 100, respectively on a 0 to 100 plastic flavor intensity scale. In successive sessions, the panel received the same 4 samples coded with 3-digit numbers and had to score them according to the consensus score, with a variation in the scale no greater than ±10 points. Once the panel was calibrated, samples were measured in triplicate using the upper scale shown in Figure 6.3. Eighteen grams of each of the 9 samples (see

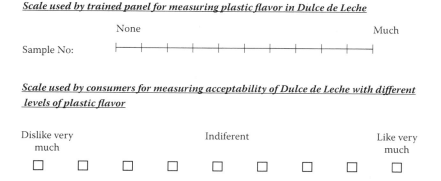

**Figure 6.3** Scales used by trained panel and consumers in determining the COP of dulce de leche.

previous section) were served in 70-ml plastic cups coded with a 3-digit number. Water and peeled slices of Granny Smith apples were provided for palate cleansing between samples.

## 6.4.4 Determination of acceptability levels of the same samples by a consumer panel

Fifty subjects between 18 and 50 years old and who consumed DL at least once a week were recruited. Section 2.7.3.1 notes that approximately 100 consumers are recommended by Hough et al. (2006) for general consumer tests; however, this article was published after the DL research was performed. For this research, a smaller group of consumers was recruited. Each consumer received the nine samples corresponding to the nine concentration levels of plastic flavor (see Section 6.4.2) and evaluated these using the lower scale shown in Figure 6.3. The samples were presented monadically in random order. Water and cracker-type biscuits were available for rinsing between samples. Following the test, the consumers received a bag with candy and chocolates as a reward for their participation. As shown in Figure 6.3, the evaluation task performed by the consumer panel was different from the task performed by the trained panel.

For DL the presentation of samples for the trained and consumer panels was similar. For other products this might not be the case. For sunflower oil (Ramírez et al. 2001) the trained panel evaluated the oil on its own and at a temperature of $50 \pm 1°C$ as recommended by the AOCS (1989). This would not be appropriate for consumers, as they would very probably give low scores to all samples regardless of whether they had an oxidized flavor. The oil samples for consumers were tasted using cubed boiled potatoes as a carrier, all at room temperature.

| Consumer | \multicolumn{9}{c}{CONCENTRATION} |
|---|---|---|---|---|---|---|---|---|---|
| | 0 | 6 | 9 | 13 | 20 | 30 | 44 | 67 | 100 |
| 1 | 8 | 9 | 8 | 9 | 7 | 8 | 4 | 2 | 3 |
| 2 | 9 | 8 | 2 | 8 | 6 | 6 | 3 | 8 | 3 |
| 3 | 7 | 4 | 6 | 2 | 3 | 2 | 2 | 1 | 1 |
| 4 | 6 | 9 | 8 | 9 | 9 | 3 | 6 | 1 | 1 |
| 5 | 9 | 9 | 9 | 8 | 9 | 9 | 7 | 4 | 5 |
| 6 | 7 | 7 | 6 | 8 | 6 | 6 | 7 | 3 | 2 |
| 7 | 5 | 8 | 7 | 5 | 6 | 1 | 1 | 5 | 1 |
| 8 | 9 | 8 | 7 | 4 | 4 | 1 | 1 | 1 | 2 |
| 9 | 8 | 9 | 5 | 8 | 6 | 4 | 1 | 3 | 1 |
| 10 | 7 | 5 | 8 | 5 | 9 | 2 | 2 | 6 | 1 |
| 11 | 9 | 3 | 7 | 4 | 6 | 4 | 1 | 2 | 1 |
| 12 | 5 | 7 | 7 | 5 | 3 | 8 | 7 | 6 | 2 |
| 13 | 2 | 9 | 7 | 3 | 6 | 7 | 1 | 9 | 1 |
| 14 | 8 | 8 | 4 | 6 | 4 | 3 | 3 | 2 | 4 |
| 15 | 8 | 8 | 8 | 4 | 3 | 7 | 4 | 2 | 2 |
| 16 | 5 | 4 | 7 | 6 | 7 | 7 | 7 | 3 | 5 |
| 17 | 7 | 6 | 6 | 6 | 4 | 5 | 2 | 4 | 1 |
| 18 | 7 | 6 | 5 | 7 | 8 | 4 | 4 | 7 | 3 |
| 19 | 8 | 8 | 8 | 8 | 8 | 8 | 5 | 4 | 3 |
| 20 | 7 | 8 | 2 | 8 | 7 | 8 | 4 | 2 | 5 |
| 21 | 7 | 7 | 4 | 4 | 6 | 4 | 4 | 4 | 6 |
| 22 | 8 | 7 | 3 | 6 | 7 | 2 | 9 | 1 | 2 |

*Figure 6.4* Acceptability data for 22 of the 50 consumers for different concentrations of plastic flavor in dulce de leche. The full data can be downloaded from the
editor's Web site: plastic_flavor.xls.

## 6.4.5 Calculation of the COP

The first step is to perform a two-way analysis of variance (ANOVA) of the
consumer data, considering the consumer and the samples as variation
factors. To be rigorous, the consumer is a random effect and the sample a
fixed effect. As each consumer measures each sample only once, whether
the consumer is considered a fixed or a random effect does not change the
final ANOVA results. This ANOVA can be done with the data analysis
tools of Excel. Figure 6.4 shows the data for 22 of the 50 consumers in form
to be analyzed by Excel:

1. If the Data Analysis command is not on the Excel Tools menu, install
   the Analysis ToolPak: on the Tools menu, click Add-Ins, select the
   Analysis ToolPak checkbox, and install it.

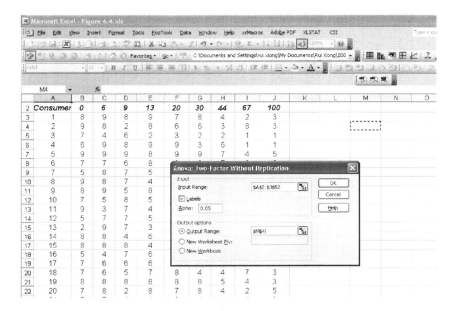

*Figure 6.5* Excel window for analysis of variance of consumer data shown in Figure 6.4.

2. Under the Data Analysis command, choose "Analysis of variance: two factors without replication." This should bring up the window shown in Figure 6.5. Here you choose the whole range of data including the labels, click on the labels box, choose a cell where you want your output to start and perform the analysis.

3. Once the data has been analyzed the parameter of interest for the COP calculation is the mean square of the error which is highlighted in Figure 6.6.

Once the mean square of the error from the ANOVA has been calculated, the following formula is applied:

$$S = F - Z_\alpha \sqrt{\frac{2MSE}{n}} \tag{6.1}$$

where

$S$     = value below which the sensory acceptability of the stored product is significantly reduced based on the 1 to 9 acceptability scale

$F$     = acceptability of fresh sample

$Z_\alpha$    = one-tailed coordinate of the normal curve for $\alpha$ significance level

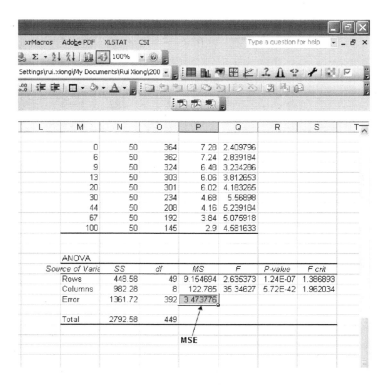

*Figure 6.6* Analysis of variance table of the dulce de leche consumer data produced by Excel. The mean square error (MSE) used to calculate the COP is highlighted.

$MSE$ = mean square of the error derived from the analysis of variance of the consumer data using consumer and sample as variation factors

$n$   = number of consumers

The coordinate of the normal curve is one-tailed because it is assumed that the product with the plastic flavor will have lower acceptability than the fresh product. Fritsch et al. (1997) and Ramírez et al. (2001) used expressions similar to Equation 6.1 in their calculations but derived from two-tailed comparisons. If α (significance level) is small, shelf life of the product will tend to increase, and if α is large, the shelf life tends to decrease. The classical 5% significance level seems to be a sensible choice; thus Z = 1.645.

For the DL plastic flavor example, Equation 6.1 results in:

$$S = 7.3 - 1.645\sqrt{\frac{2 \times 3.474}{50}} = 6.7$$

The second step in calculating the COP is to perform a regression of the acceptability data (Section 6.4.4) versus the trained sensory panel data (Section 6.4.3). The data to be regressed are in Table 6.1, and the resulting linear regression is presented in Figure 6.7. As observed in Figure 6.7, by introducing the value of S = 6.7, a sensory COP of 24 on the 0 to 100 sensory scale was obtained. The interpretation of this COP is that when a

*Table 6.1* Consumer Acceptability and Trained Sensory Panel Average Scores for Samples with Increasing Intensity of Plastic Flavor in Dulce de Leche

| % of stock reference[a] | Consumer acceptability[b] | Trained sensory panel[c] |
|---|---|---|
| 0 | 7.28 | 8.95 |
| 6 | 7.24 | 25.18 |
| 9 | 6.48 | 21.37 |
| 13 | 6.06 | 33.42 |
| 20 | 6.02 | 40.07 |
| 30 | 4.68 | 68.53 |
| 44 | 4.16 | 72.96 |
| 67 | 3.84 | 65.78 |
| 100 | 2.9 | 79.43 |

[a] Stock reference: 250 g of DL in its polystyrene pot heated at 80°C for 24 hours.
[b] Measured on a 1 to 9 hedonic scale.
[c] Measured on a 0 to 100 sensory intensity scale.

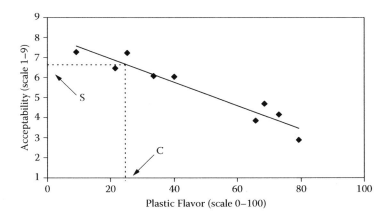

*Figure 6.7* Sensory acceptability scores versus sensory panel scores for dulce de leche with different levels of plastic flavor. S = minimum tolerable acceptability of stored sample, C = sensory failure COP. (Reprinted with permission from: Garitta, L., G. Hough, and R. Sánchez. 2004. Sensory shelf life of dulce de leche. *Journal of Dairy Science* 87: 1601–1607.)

panel trained with the references as described in Section 6.4.3 measures a sample of DL and the average score is ≥24, the average consumer will have reduced his acceptability in relation to the fresh sample.

The correlation between acceptability scores and trained sensory panel scores does not necessarily have to be linear. Hough et al. (2002), in their study on COPs of powdered milk, used linear, exponential, and logistic regressions. Figure 6.8 shows acceptability versus trained panel scores for the lypolysis defect with the straight line and logistic regressions. The logistic equation gave a slightly better fit than the linear equation, but it can be observed that the sensory failure COP was similar using either equation. For oxidized flavor (not shown) the exponential regression gave a slightly better fit than the straight line, but the COPs were similar. The inherent variability in sensory data results in few practical differences when applying one regression model or another. For most situations the straight line regression is adequate.

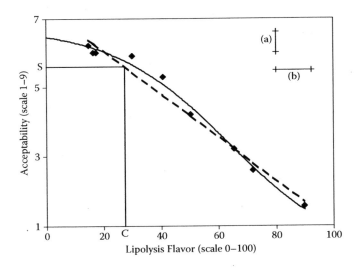

*Figure 6.8* Sensory acceptability of Argentine consumers versus trained sensory panel scores for reconstituted milk powder with different levels of lipolysis flavor. Curve corresponds to logistic equation and straight line to linear equation. S = value below which the sensory acceptability is significantly reduced, C = sensory failure COP. (a) least significant difference from acceptability ANOVA, (b) least significant difference from trained panel ANOVA. (Reprinted with permission from: Hough, G., R.H. Sánchez, G. Garbarini de Pablo, R.G. Sánchez, S. Calderón Villaplana, A.M. Giménez, and A. Gámbaro. 2002. Consumer acceptability versus trained sensory panel scores of powdered milk shelf-life defects. *Journal of Dairy Science* 85: 2075–2080.)

## 6.5 Introduction to kinetics

Figure 6.1 shows how oxidized flavor changes versus storage time for sunflower oil stored at 45°C for 90 days. In this case a straight line regression was calculated from the experimental points. As we shall see this implies zero-order kinetics. However zero-order kinetics doesn't necessarily have to be the case. In the present section the equations based on zero-order and first-order kinetics will be presented.

### 6.5.1 Zero-order kinetics

In a reaction following zero-order kinetics, the reaction rate is constant:

$$Reaction\ rate = \frac{dS}{dt} = k \qquad (6.2)$$

where
$S$ = level of a sensory defect, for example oxidized flavor or brown color,
$t$ = storage time, and $k$ = reaction rate are constant.

If Equation 6.2 is integrated:

$$S = S_0 + k \cdot t \qquad (6.3)$$

where $S_0$ = level of sensory defect at time = 0.
  Equation 6.3 is for a sensory defect that increases over time. If the critical descriptor decreases over time, the plus sign in Equation 6.3 changes to minus. For example, if the loss of crispness in a snack follows zero-order kinetics, the equation would be:

$$Crispness = Crispness_0 - k \cdot t \qquad (6.4)$$

Both Equations 6.3 and 6.4 represent straight lines, with positive and negative slopes, respectively. Figure 6.1 shows an example of a straight-line relationship between oxidized flavor versus storage time.

### 6.5.2 First-order kinetics

In a reaction following first-order kinetics, the reaction rate is a function of the measured parameter:

$$Reaction\ rate = \frac{dS}{dt} = k \cdot S \qquad (6.5)$$

If Equation 6.5 is integrated:

$$\ln S = \ln S_0 + k \cdot t \tag{6.6}$$

Equation 6.6 is for a sensory defect that increases with storage time. For a sensory property that decreases with storage time a similar equation is obtained:

$$\ln Crispness = \ln Crispness_0 - k \cdot t \tag{6.7}$$

If a logarithmic scale is used for the y-axis, Equations 6.6 and 6.7 also represent straight lines with positive and negative slopes, respectively.

### 6.5.3   Choosing between zero- and first-order kinetics

The classical method to choose between zero- and first-order kinetics is to graph Equations 6.3 and 6.6 if measuring an increasing sensory defect, or Equations 6.4 and 6.7 if measuring a decreasing sensory property. The graph where the experimental points best approximate a straight line would correspond to the most appropriate reaction rate. However, with sensory data, which generally do not have high precision, this procedure is not so straightforward, as shown in the following example.

Figure 6.9 represents the development of acid flavor in fat-free stirred vanilla yogurt measured by a trained sensory panel. The visual comparison of the experimental points with the regression lines does not conclusively favor any one of the kinetic orders; neither do the $R^2$ values which are similar. The sum of squared differences between experimental and predicted values was 524 and 451 for zero-order and first-order kinetics, respectively. These values are of similar magnitude. Development of acid flavor in yogurt is not the result of a single chemical reaction following a specific kinetic order; rather, it is the result of several reactions all leading to an increase of acid flavor during storage. For example, there is the production of lactic acid by microbial fermentation and the decrease in sweetness due to non-nutritive sweetener's molecules breaking down; both reactions contribute to an increase in acid flavor. The overall kinetic order does not necessarily have to be strictly zero-order or first-order.

As pointed out by Taoukis et al. (1997), when the reaction is not carried far enough (less than 50% conversion) both zero- and first-order might be indistinguishable. Additionally, the worse the precision of the measuring method, the larger the extent of change to which the experiment should be carried out to obtain an acceptably accurate estimate of the reaction rate constant.

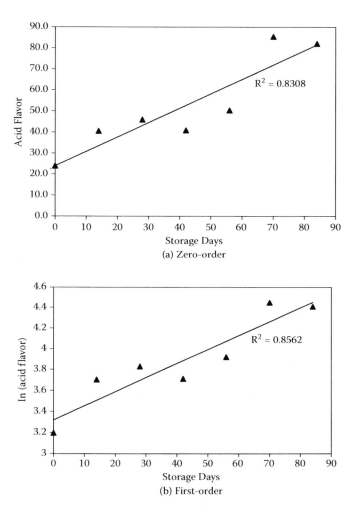

*Figure 6.9* Development of acid flavor for fat-free stirred vanilla yogurt measured by a trained sensory panel. (a) Zero-order kinetics, (b) first-order kinetics.

## 6.5.4 Sensory properties that present a lag phase

Some sensory properties present a lag phase in their development during storage. During a typical lag period there is a buildup of a critical intermediate concentration. The rate of the reaction during the buildup period is normally slower.

Curia and Hough (2009) studied the development of dark color during the storage of a human milk replacement formula. Figure 6.10 shows sensory color changes versus storage time; first there was a lag phase

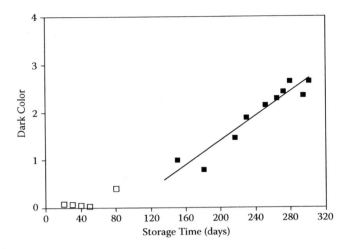

**Figure 6.10** Dark color development in a human milk replacement formula as measured by a trained sensory panel. Clear boxes represent the lag phase and dark boxes the increasing phase of the dark color development.

time, and then there was a linear increasing phase. This lag phase has been observed for the Maillard reaction in other systems (Buera et al. 1987, 1986). Curia and Hough (2009) considered the lag phase to last as long as the color was below 0.5 on the 0 to 10 sensory scale. During the lag phase there was no color formation; therefore, the regression for the increasing phase was used. This increasing phase was linear, indicating a zero reaction order.

## 6.6    Using the COP to estimate shelf life

The COP indicates a limit on an analytical sensory scale or an instrumental measurement above which (or below which, for a desirable parameter) the consumer's acceptability of the product significantly decreased. This value is the one to be used in establishing the SSL once the evolution of the critical descriptor or instrumental parameter has been established as a function of storage time.

### 6.6.1    Sample storage and trained sensory panel evaluations

The SSL of DL (Garitta et al. 2004) based on the development of plastic flavor as the critical descriptor will be used to illustrate the procedure. Commercial, home-use DL provided by a leading manufacturer in Argentina was used. The product was all from the same batch, packaged in 250-g polystyrene pots. This is standard packaging for DL in Argentina and other countries. DL is stable at room temperature. The pots were stored at 5 ± 1°C until use; it was considered that changes at this temperature were negligible compared with changes under storage at the target temperature of 37°C. A reversed storage design was used (see Section 3.3.6.2) whereby at different times samples were removed from 5°C storage and placed at 37°C storage approximately every 10 days for a total of 122 days. Thus, after 122 days, samples with different storage times at 37°C were available for sensory evaluation by the trained panel. For plastic flavor, samples were taken from the DL in contact with the polystyrene pot. Consumers usually scrape the inside of the DL pot with a spoon, so this would be the critical portion of DL.

To train the nine-member panel, the plastic flavor defect was prepared as indicated in Section 6.4.2, that is, by heating 250 g of DL in its polystyrene pot for 24 hours at 80°C. Every 6 hours the DL was stirred in the pot for approximately 2 minutes. This stock reference was diluted in untreated DL to obtain different intensities of the defect; the following percent stock references were used to prepare references to train the panel in the plastic flavor scale: 0, 6, 20, and 100%. These concentrations corresponded to 0, 10, 50, and 100 on the plastic flavor 0–100 sensory scale, respectively. In successive sessions, the panel received these four samples coded with 3-digit numbers and had to score them according to the consensus score, with a variation in the scale no greater than ±10 points.

Once the panel was calibrated, samples stored at 37°C were measured in duplicate; 18 g of each of the samples were served in 70-ml plastic cups coded with a 3-digit number. Water and peeled slices of Granny Smith apples were provided for palate cleansing between samples.

## 6.6.2 Results and calculations

The results of the development of plastic flavor during storage of DL at 37°C are presented in Table 6.2. These same results are graphed in Figure 6.11. A straight-line regression was calculated supposing a zero-order reaction rate. Confidence intervals of the regression line were calculated using the following formula (Drapper and Smith 1981):

$$P_0 \pm (\upsilon, 1-\alpha/2) \left\{ \frac{1}{n} + \frac{\left(t_0 - t_{mean}\right)^2}{\sum \left(t_i - t_{mean}\right)^2} \right\}^{1/2} s \qquad (6.8)$$

where

$P_0$      = predicted value of plastic flavor for storage time $t_0$

$n$      = number of experimental points from which the regression line was calculated

*Table 6.2* Trained Sensory Panel Average
Scores for Samples with Increasing Intensity of
Plastic Flavor in Dulce de Leche Stored at 37°C

| Storage time (days) | Plastic flavor[a] |
|:---:|:---:|
| 0 | 0 |
| 4 | 1 |
| 9 | 4 |
| 15 | 2 |
| 22 | 12 |
| 30 | 15 |
| 39 | 18 |
| 49 | 20 |
| 60 | 18 |
| 66 | 34 |
| 108 | 42 |
| 122 | 75 |

[a] Plastic flavor measured by a nine-member trained
panel by duplicate on a 0 to 100 sensory scale.

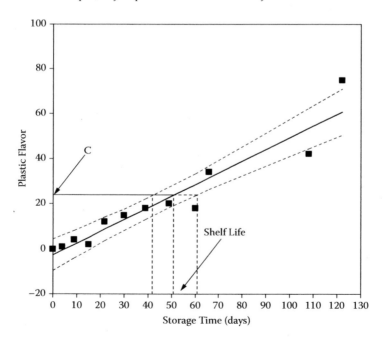

*Figure 6.11* Experimental points with straight line regression and 95% confidence
bands for plastic flavor versus storage time for dulce de leche stored at 37°C. Shelf
life with corresponding 95% confidence interval is indicated. C: COP.

$t(\upsilon, 1-\alpha/2)$ = Student $t$-value for $\upsilon = n - 2$ degrees of freedom and an $\alpha$ significance level

$t_0$ = storage time for which the prediction is wanted

$t_{mean}$ = mean value of experimental storage times

$s$ = standard error of the regression

This last value can be obtained from the Summary Output after fitting a linear regression with Excel's Data Analysis menu. The value of s can be read as the *Standard error* under the *Regression Statistics*. Alternatively, s can also be calculated as the square root of the error term in the regression analysis of variance.

The linear regression obtained from the data presented in Table 6.2 for plastic flavor in DL was:

$$\text{Plastic} = -2.58 + 0.519 \times \text{Storage time} \tag{6.9}$$

To illustrate the use of Equation 6.8, a value of $t_0 = 30$ days will be taken:

$$12.99 \pm 2.228 \times \left\{ \frac{1}{12} + \frac{186.8}{17,250.7} \right\}^{1/2} \times 7.05 = 12.99 \pm 4.82$$

These confidence intervals can be calculated for multiple values of storage time, and with these, the confidence bands of the linear regression can be drawn as shown in Figure 6.11. Once this stage has been reached, the shelf life can be estimated by introducing the COP in the linear regression equation. In Section 6.4.5 the COP for plastic flavor in DL was estimated to be = 24 on the 0 to 100 sensory scale; using Equation 6.9 this results in an estimated shelf life at 37°C of:

$$\textit{Shelf life} = \frac{24 + 2.58}{0.519} = 51 \text{ days}$$

Confidence limits can be estimated as shown in Figure 6.11; that is where the upper and lower confidence bands of the regression equal the COP of 24. A simple way to obtain these values in Excel is to copy a series of increasing storage-time values in one column and the formula given by Equation 6.8 in adjacent columns (one column for the upper bound and another column for the lower bound). Where these last two columns equal approximately 24, that is, the COP, the confidence limits under the storage time column can be read. For our DL example the 95% lower and upper confidence bands were 42 days and 61 days, respectively.

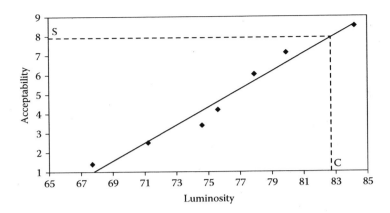

*Figure 6.12* Acceptability of a human milk replacement formula versus CIE-LAB luminosity value. S = value below which the sensory acceptability is significantly reduced, C = instrumental COP.

## 6.7　Instrumental COPs

In the above example (Section 6.4.5) the COP was determined on plastic flavor, a measurement performed by a trained sensory panel. If there is a strong correlation between loss of acceptability and an instrumental measurement, then it can be convenient to determine the COP on an instrumental scale. Curia and Hough (2009) studied the development of dark color during the storage of a human milk replacement formula. They found a high correlation between sensory dark color measured by a trained panel and CIE-LAB luminosity values (L*). This allowed determining an L* COP C = 82.7 as shown in Figure 6.12. The interpretation of this COP is that when the L* value is below 82.7 (product gets darker), the acceptability of the stored product is significantly reduced in relation to the fresh product. The possibility of establishing an instrumental COP is dependent on finding a high correlation between instrumental and sensory parameters.

## 6.8　Caveats for using COP methodology

In Section 6.1 the COP methodology was presented as an alternative when the survival analysis method is difficult to apply. When applying the COP methodology, the following caveats should be considered:

1. Equation 6.1 defines the value of S = value below which the sensory acceptability of the stored product is significantly reduced based on the 1 to 9 acceptability scale. This value of S is then used to estimate the COP (Section 6.4.5). For the example of DL, this was the plastic

flavor above which the acceptability of consumers was significantly reduced. This reduction in acceptability does not necessarily mean that the product will be rejected. The consumer may reason, "The product has a slightly lower acceptability, but I still don't mind eating it." Thus estimations based on the COP methodology would tend to be conservative. Giménez et al. (2007) reached this conclusion when comparing the COP and survival analysis methodologies in estimating the shelf life of brown pan bread.

2. It is sensitive to the chosen critical descriptor. Hough et al. (2002) estimated COPs for different defects in reconstituted milk powder. For dark color, the percentage of consumers who would reject the sample at the sensory failure COP was only 4%, compared with approximately 30% for the flavor descriptors. Acceptability of dark-colored samples close to the fresh sample was reduced significantly, yet few consumers rejected these samples. If color were chosen as the critical descriptor, an erroneously short shelf life would be estimated.

3. It is sensitive to the number of consumers. This is mainly because the formula for calculating S = value below which the sensory acceptability of the stored product is significantly reduced contains n = number of consumers (Equation 6.1). In the above example of plastic flavor in DL, suppose that 30 consumers were recruited instead of the 50 reported above (Section 6.4.4). Choosing 30 randomly from the original 50 gave the following results:

Mean square error = 3.48
S = 6.51
COP = 26
Estimated sensory shelf life = 46 days

This last value is less than the 51 days estimated for the 50 consumers. As mentioned above (Section 6.4.4) the recommended number of consumers is approximately 100.

## *References*

AOCS. 1989. Flavor panel evaluation of vegetable oils. AOCS Recommended Practice Cg 2–83. Champaign, Illinois: American Oil Chemists Society.

Buera, M.P., R.D. Lozano, and C. Petriella. 1986. Definition of color in the non-enzymatic browning process. *Die Farbe* 32/33: 316–326.

Buera, M.P., J. Chirife, S.L. Resnik, and G. Wetzler. 1987. Non-enzymatic browning in liquid model systems of high water activity. 2. Kinetics of color changes due to Maillard reaction between single sugars and glycine and comparison with caramelization browning. *Journal of Food Science* 52: 1063–1067.

Curia, A.V., and G. Hough. 2009. Selection of a sensory marker to predict the sensory shelf life of a fluid human milk replacement formula. *Journal of Food Quality* 32: 793–809.

Drapper, N. R., and H. Smith. 1981. *Applied regression analysis*, Chapter 1. New York: John Wiley & Sons.

Fritsch, C. W., C. N. Hofland, and Z. M. Vickers. 1997. Shelf life of sunflower kernels. *Journal of Food Science* 62: 425–428.

Garitta, L., G. Hough, and R. Sánchez. 2004. Sensory shelf life of dulce de leche. *Journal of Dairy Science* 87: 1601–1607.

Giménez, A., P. Varela, A. Salvador, G. Ares, S. Fiszman, and L. Garitta. 2007. Shelf life estimation of brown pan bread: A consumer approach. *Food Quality and Preference* 18: 196–204.

Gómez-Guillén, M.C., M.E. López-Caballero, O. Martínez-Alvarez, and P. Montero. 2007. Sensory analyses of Norway lobster treated with different antimelanosis agents. *Journal of Sensory Studies* 22: 609–622.

Hough, G., M.L. Calle, C. Serrat, and A. Curia. 2007. Number of consumers necessary for shelf-life estimations based on survival analysis statistics. *Food Quality and Preference* 18: 771–775.

Hough, G., R.H. Sánchez, G. Garbarini de Pablo, R.G. Sánchez, S. Calderón Villaplana, A.M. Gimenez, and A. Gámbaro. 2002. Consumer acceptability versus trained sensory panel scores of powdered milk shelf-life defects. *Journal of Dairy Science* 85: 2075–2080.

Hough, G., I. Wakeling, A. Mucci, E. Chambers IV, I. Méndez Gallardo, and L. Rangel Alves. 2006. Number of consumers necessary for sensory acceptability tests. *Food Quality and Preference* 17: 522–526.

Moro, O., and G. Hough. 1985. Total solids and density measurements of dulce de leche, a typical Argentine dairy product. *Journal of Dairy Science* 68:521–525.

Muñoz A., G.V. Civille, and B.T. Carr. 1992. *Sensory evaluation in quality control*, pp. 80–81. New York: Van Nostrand Reinhold.

Nunes, M.C., J.P. Emond, and J.K. Brecht. 2006. Brief deviations from set point temperatures during normal airport handling operations negatively affect the quality of papaya (*Carica papaya*) fruit. *Postharvest Biology and Technology* 41: 328–340.

Patsias, A., I. Chouliara, A. Badeka, I.N. Savvaidis, and M.G. Kontominas. 2006. Shelf-life of a chilled precooked chicken product stored in air and under modified atmospheres: Microbiological, chemical, sensory attributes. *Food Microbiology* 23: 423–429.

Ramírez, G., G. Hough, and A. Contarini. 2001. Influence of temperature and light exposure on sensory shelf life of a commercial sunflower oil. *Journal of Food Quality* 24: 195–204.

Taoukis, P., T.P. Labuza, and I.S. Saguy. 1997. Kinetics of food deterioration and shelf-life prediction. Chapter 10 in *Handbook of food engineering practice*, ed. K. Valentas, E. Rotstein, and R.P. Singh. Boca Raton, FL: CRC Press.

# chapter 7

# Accelerated storage

## 7.1  Introduction

Today's food producers face strong pressure to develop new products in record time, while improving productivity, shelf life, and overall quality. Estimating the rejection-time distribution for products that have a relatively long shelf life is particularly complicated. For example, the design, development, and marketing of a new type of biscuit may allow only 6 weeks, while the SSL of the biscuit will be in the range of 6 to 9 months. In such cases accelerated storage tests are used to hopefully obtain long-term estimations based on short-term tests.

Section 3.1 mentioned that (a) foods are heterogeneous, (b) the chemical reactions involved in the deterioration of foods have complex kinetics, (c) rheological and textural properties are not simply modeled, (d) biological changes continue during food storage, and (e) the deterioration of a food during storage is often the result of a number of simultaneous processes. All this leads to difficult practical and statistical issues involved in accelerating the lives of complex products like most foods. The usual procedure is for information from tests at high levels of one or more accelerating variables (e.g., temperature, moisture, or light) to be extrapolated to obtain estimates of shelf life at lower, normal levels of the accelerating variable. Such testing saves much time and money. Pitfalls of accelerated tests will be discussed in Section 7.5.

### 7.1.1  Acceleration factor fallacy

Some food companies use, or aspire to use, a single acceleration test condition. Shelf-life estimation based on this single condition depends on certain assumptions. Often the assumptions are poorly satisfied, and the estimates may be quite crude (Nelson 1990). A typical example would be to test a powdered soup at 40°C; if the SSL at this accelerated temperature is 2 months, then an acceleration factor of 4 is applied to estimate SSL at 20°C to be 4 × 2 months = 8 months. In some companies this acceleration factor of 4 between storage at 40°C and storage at 20°C is considered valid for all products and formulations. Section 1.5.5, when addressing Taub and Singh's book (1998), pointed out that Yang (1998) introduced a conceptual error when he stated that "accelerated testing at 38°C is based on the principle that reactions are reduced half by each drop of 10°C in temperature." This rule, based on the validity of a single activation energy valid for all food products, is false. The assumptions when using these acceleration factors are basically the following (Nelson 1990):

1. *Known factor.* It is usually assumed that the acceleration factor is known. However, its value is often a company tradition of unknown origins. Even if it was once measured, its application does not take into account the uncertainty in its value due to randomness in the statistical sample of the test. A larger error entails if this factor was based on different formulations along with the error of applying the factor to completely different food products.
2. *Same σ.* In Section 4.7 log-normal and Weibull distributions were presented. These are characterized by the μ and σ parameters. When multiplying test data by an acceleration factor a constant σ value for different values of the accelerating factor is assumed. Although this is generally a reasonable assumption, it is not always so.
3. *Failure modes.* Often there is more than one failure mode and it is assumed that they all respond to the same acceleration factor. Typically, different failure modes have different acceleration factors. For example, if biscuits are stored at a high temperature, package permeability could lead to loss of crispness as the critical descriptor. The critical descriptor for the same biscuits at room temperature could be oxidized flavor. Food products, as mentioned above, are complex heterogeneous systems, and to assume a single acceleration factor is risky.

In my experience when teaching SSL courses it is often within participants' expectations to return from the course with a single acceleration factor to be applied to all their production line. It is clear that this aspiration is a fallacy.

## 7.1.2   Methods of acceleration

For general shelf-life and reliability tests of different materials and products, there are two different methods of accelerating deterioration (Meeker and Escobar 1998):

- *Increase the use rate of the product.* This method is used mainly for mechanical or electrical products. For example, if a toaster is designed to be turned on and off twice a day, the failure of the toaster can be accelerated if the toaster is turned on and off every minute, that is, 144 times per day. This method to accelerate failure does not apply to foods.

- *Increase the aging rate of the product.* This is typically done for food products by increasing the impacts of experimental storage variables. The most common accelerating factor is temperature, which will be covered in the present chapter; however, other accelerating factors such as humidity or light can be used. Humidity has received little attention as an acceleration factor, especially in relation to sensory shelf-life studies (Mizrahi and Karel 1977a, 1977b; Fu and Labuza 1993). Light was used by Ramírez et al. (2001) as an acceleration factor in storage of sunflower oil.

## 7.2 Arrhenius equation and activation energy

### 7.2.1 Arrhenius equation

In Section 6.5.1 zero-order kinetics were presented whereby the change in a sensory descriptor with storage time was described by the following equation:

$$S = S_0 + k \cdot t \tag{7.1}$$

where

$S$ = level of a sensory defect, for example oxidized flavor or brown color
$S_0$ = level of sensory defect at time = 0
$t$ = storage time
$k$ = reaction rate constant

Equation 7.1 is valid for isothermal conditions. The most widely used model for describing the relationship between $k$, the reaction rate, and temperature, is the Arrhenius equation:

$$k = k_0 \exp\left(-\frac{E_a}{RT}\right)$$

A more convenient way of expressing this equation is by introducing a reference temperature:

$$k = k_{ref} \exp\left(-\frac{E_a}{R}\left(\frac{1}{T} - \frac{1}{T_{ref}}\right)\right) \tag{7.2}$$

where

$k_{ref}$ = reaction rate constant at the chosen $T_{ref}$ ('sensory scale'/time)
$E_a$ = activation energy (cal/mol)
$R$ = gas law constant (1.98 mol. K/cal)

$T$    = temperature at which k is to be estimated (K)
$T_{ref}$ = reference temperature (K)

$T_{ref}$ is chosen within the range of considered temperatures. For example, for an accelerated study on mayonnaise where tests are to be performed at 25°C, 35°C, and 45°C, an appropriate value for $T_{ref}$ would be 300 K, equivalent to 27°C.

Equation 7.2 can be linearized by applying logarithms:

$$\ln k = \ln k_{ref} - \frac{E_a}{R}\left(\frac{1}{T} - \frac{1}{T_{ref}}\right) \tag{7.3}$$

thus by plotting ln(k) versus $(1/T - 1/T_{ref})$ will produce a straight line whose slope is $-E_a/R$.

It is to be understood that the Arrhenius equation used to model changes in reaction rates with temperature is a purely empirical relationship. Other relationships have been used, like the Eyring relationship (Meeker and Escobar 1998). When zero-order kinetics prevail, as is the case of the majority of sensory changes with storage time, the Eyring relationship presents little difference to the more simple Arrhenius equation. This Arrhenius equation is preferred as $E_a$, the activation energy, is widely used as a reference parameter to quantify the change in reaction rates with temperature.

### 7.2.2   Data for activation energy calculations

To illustrate activation energy $(E_a)$ calculations data from a mayonnaise shelf-life experiment will be used (Martínez et al. 1998). Commercial mayonnaise was stored at 20°C, 35°C, and 45°C. A control sample was stored at 5°C; it was considered that at this temperature, changes were negligible in comparison to changes at temperatures ≥20°C. A seven-member trained sensory panel compared the stored samples versus the control using quantitative descriptive analysis (Stone and Sidel 2004). For each descriptor a 12-cm non-structured scale was used, anchored at the center with *equal to control*, at the extreme left with *a lot less than control*, and at the extreme right with *a lot more than control*. The critical descriptor was oxidized flavor, which increased with storage temperature; but as other descriptors were also followed, some of which might have decreased with storage time, all descriptors were measured using the same bipolar scale.

Table 7.1 shows the results. Each oxidized flavor value shown in the table is the result of averaging over the seven assessors by duplicate, that is, it represents the average of 14 numbers. Note that total storage time for each temperature differed. If the maximum storage time had been the

***Table 7.1*** Oxidized Flavor of Mayonnaise Stored
at Different Temperatures and Different Times

| Storage temperature | | | | | |
|---|---|---|---|---|---|
| 20°C | | 35°C | | 45°C | |
| Days | Oxidized[a] | Days | Oxidized | Days | Oxidized |
| 122 | 4 | 11 | 8.1 | 7 | 2.3 |
| 145 | 8.1 | 20 | 13.9 | 14 | 7.5 |
| 164 | 6.6 | 30 | 19 | 21 | 15.3 |
| 183 | 16.9 | 39 | 23 | 28 | 24.4 |
| 201 | 19.3 | 48 | 31.6 | 35 | 36.9 |
| 224 | 21.2 | 56 | 33.2 | 42 | 43.4 |
| 245 | 28.2 | 61 | 37.2 | | |

[a] Oxidized flavor is on a 0 to 60 sensory scale.

same for all temperatures, for example, 42 days, then the samples stored
at 20°C would have shown no appreciable change.

## 7.2.3 Simple activation energy calculations

Equation 7.3 shows that plotting $\ln(k)$ versus $(1/T - 1/T_{ref})$ will produce a
straight line whose slope is $-E_a/R$. So the first step with this methodology
is to obtain the reaction rate constants at each temperature. Figure 7.1
shows oxidized flavor versus storage time for mayonnaise stored at 35°C.
The linear correlation indicates that zero-order kinetics was appropriate.
The slope of the straight line is the reaction rate constant corresponding to
this temperature. The linear correlation of oxidized flavor versus storage

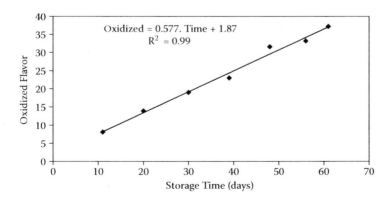

***Figure 7.1*** Oxidized flavor versus storage time for commercial mayonnaise stored
at 35°C.

**Table 7.2** Reaction Rate Constants for Oxidized Flavor Development in Mayonnaise Corresponding to Different Storage Temperatures

| Temperature (°C) | Reaction rate constant (k, oxidized flavor/days) | (1/T°K–1/300°K) | ln(k) |
|---|---|---|---|
| 20 | 0.1972 | 7.964E–05 | –1.623 |
| 35 | 0.577 | –8.658E–05 | –0.550 |
| 45 | 1.2359 | –18.87E–05 | 0.212 |

time is repeated for each one of the tested temperatures obtaining the values shown in Table 7.2. To estimate the activation energy the regression of $\ln(k)$ versus $(1/T - 1/T_{ref})$ is calculated, that is, the fourth versus the third column of Table 7.2. In Excel there are two ways of obtaining the regression equation:

1. Plot $\ln(k)$ versus $(1/T - 1/T_{ref})$, and on this plot right-click on one of the experimental points and choose the Add Trendline tab. Within the options you can check the Display Equation on Chart box, and this will show the regression equation on the plot; the slope of this equation is the estimated value of $-E_a/R$. This estimated value is correct; however, confidence intervals for the slope are not shown.
2. Use the Regression Data Analysis tool. If the Data Analysis command is not on the Excel Tools menu, install the Analysis ToolPak: on the Tools menu, click Add-Ins, select the Analysis ToolPak checkbox, and install it. Under the Data Analysis command, choose Regression. This should bring up the window shown in Figure 7.2. Here you choose the $(1/T - 1/T_{ref})$ column as the X Range and $\ln(k)$ as the Y Range. If you include the column labels, click on the labels box, choose a cell where you want your output to start, and perform the analysis. Once the data have been analyzed, the intercept and slope of the regression can be read from the coefficient table as shown in Figure 7.3. This coefficient table also shows the confidence intervals of the parameter estimates (Excel unnecessarily prints these twice). For the present mayonnaise data the following equation results:

$$\ln k = \ln k_{ref} - \frac{E_a}{R}\left(\frac{1}{T} - \frac{1}{T_{ref}}\right) = -1.097 - 6803\left(\frac{1}{T} - \frac{1}{300}\right)$$

The $E_a/R \pm 95\%$ confidence interval is $6803 \pm 3389$; and the $E_a \pm 95\%$ confidence interval is $13470 \pm 6710$ cal/mol.

When performing the regressions by either of the above methods (a) or (b), the correlation coefficient = 0.99. This is a general measure used to test the fit of a regression. However, when there are a reduced number of

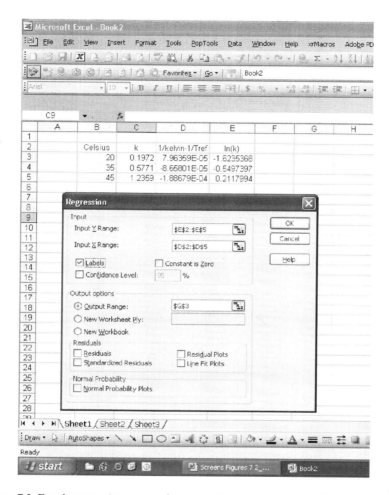

*Figure 7.2* Excel regression menu for activation energy calculation of data presented in Table 7.2.

experimental points, this coefficient is of little value. It is clear that the uncertainty in the estimated activation energy is high, as reflected by the high value for the confidence interval. This uncertainty is mainly due to the small number of experimental points, three in this example. To obtain a more reliable estimate of the activation energy by this simple linear regression method, the only alternative is to store and evaluate samples at a minimum of six different temperatures. The experimental work involved in this alternative is usually prohibitive in time and costs.

Many published articles report activation energy values based on a limited number of experimental points without including confidence

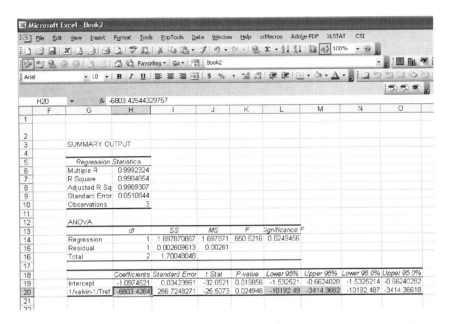

***Figure 7.3*** Excel output from regression analysis for activation energy calculation of data presented in Table 7.2.

intervals. For example, Almeida-Domínguez et al. (1992) published activation energy values for loss of sensory acceptability in stored snacks based on only two temperature points with no confidence intervals reported. In another example, Sithole et al. (2005) reported activation energy values for the development of brown color in whey powder using three temperature points; no confidence intervals were reported.

### 7.2.4   *Activation energy calculations based on non-linear regression*

Figure 7.1 showed that oxidized flavor increase for mayonnaise stored at 35°C followed zero-order kinetics:

$$Ox = Ox_0 + k \cdot t \tag{7.4}$$

The reaction rate constant in Equation 7.4 can be replaced by its value as a function of temperature given by Equation 7.2 to obtain:

$$Ox = Ox_0 + k_{ref} \exp\left(-\frac{E_a}{R}\left(\frac{1}{T} - \frac{1}{T_{ref}}\right)\right) \cdot t \tag{7.5}$$

Supposing first-order kinetics were to be adopted:

$$\ln Ox = \ln Ox_0 + k \cdot t$$

And Equation 7.5 for first order kinetics would be:

$$\ln Ox = \ln Ox_0 + k_{ref} \exp\left(-\frac{E_a}{R}\left(\frac{1}{T} - \frac{1}{T_{ref}}\right)\right) \cdot t \tag{7.6}$$

Equations 7.5 and 7.6 have three unknown parameters: $Ox_0$ ($\ln Ox_0$ for first-order kinetics), $k_{ref}$ and $E_a/R$; this last parameter is non-linear. Ordinary least squares regression cannot be used to estimate the value of these parameters that best fit the experimental points. Non-linear regression routines have to be used. The first step in using these routines is to provide them with initial approximations of the parameters to be estimated:

- $Ox_0$: This is the average oxidized flavor for the fresh product. In some case there may even be an experimental estimate. From Table 7.1 an approximate $Ox_{0\_initial} = 1$. If first-order kinetics is chosen, then $Ox_{0\_initial}$ should never be taken $= 0$ as the $\ln(0)$ (Equation 7.6) is undetermined.
- $k_{ref\_initial}$: This is the reaction rate constant at the chosen reference temperature. The reference temperature is chosen within the ranges of the accelerated storage design. Thus for the mayonnaise example, where measurements were at 20°C, 35°C, and 45°C, $T_{ref}$ was chosen to be 27°C = 300 K. For one of the temperatures close to $T_{ref}$, we can perform a linear regression to have an estimate of $k_{ref\_initial}$. For our example, a linear regression using zero-order reaction for temperature = 35°C, gave $k_{35°C} = 0.577$ (see Table 7.2); thus, we can assume $k_{refinitial} = 0.5$. For the mayonnaise example, a linear regression assuming first-order kinetics for temperature = 35°C, gave $k_{35°C} = 0.029$; thus, we can assume $k_{ref\_initial} = 0.02$.
- $E_a/R_{initial}$: This is an initial value for the activation energy expressed in cal/mol divided by $R$ (the gas law constant = 1.98 for these units). Activation energies are usually within the range of 5,000 to 30,000 cal/mol; thus, $E_a/R$ would be between 2,500 and 15,000. An initial value of 10,000 is usually adequate.

When estimating $E_a$ by simple linear regression as shown in Section 7.2.3, the complete data set of 20 experimental points (Table 7.1) was first reduced to three experimental points (Table 7.2), and with this reduced set of data $E_a$ was calculated; as only three points were used in this final estimation, confidence intervals were wide. The basic idea behind the non-linear estimation of the activation energy is to make use of the complete data set in the estimation.

A number of commercial statistical software packages, such as Genstat (VSN International Inc., Hemel Hempstead, United Kingdom) that also has the Genstat Discovery version of free access for academic institutions in the developing world, have procedures for non-linear regression estimations. R, a free-access statistical package (http://www.r-project.org/, accessed May 26, 2009) also has a procedure for non-linear regression estimations and is used in the present book.

The first step in performing calculations with R is to have the raw data in an Excel spreadsheet in the format shown in Table 7.3. For the data to be compatible with the function written in R for $E_a$ non-linear estimations, it should have the following characteristics: The first column should indicate the storage time; the second column, the storage temperature in K; and the third column, the measured parameter. In this case the measured parameter was oxidized flavor, but it could be any

*Table 7.3* Oxidized Flavor of Mayonnaise Stored at Different Temperatures and Different Times in a Format to Be Read by R

| Days of Storage | Kelvin Units | Oxidized Flavor |
|:---:|:---:|:---:|
| 122 | 293 | 4 |
| 145 | 293 | 8.1 |
| 164 | 293 | 6.6 |
| 183 | 293 | 17 |
| 201 | 293 | 19 |
| 224 | 293 | 21 |
| 245 | 293 | 28 |
| 11 | 308 | 8.1 |
| 20 | 308 | 14 |
| 30 | 308 | 19 |
| 39 | 308 | 23 |
| 48 | 308 | 32 |
| 56 | 308 | 33 |
| 61 | 308 | 37 |
| 7 | 318 | 2.3 |
| 14 | 318 | 7.5 |
| 21 | 318 | 15 |
| 28 | 318 | 24 |
| 35 | 318 | 37 |
| 42 | 318 | 43 |

other sensory or physicochemical parameter. All columns should have text headings.

The data in the format shown in Table 7.3 should be saved as a tab-delimited text file (extension txt). This format can be easily read by *R*. Once *R* has been installed, the instructions to analyze the mayonnaise data saved in the text file are as follows:

1. Open R.
2. Change working directory (if necessary).
3. Read raw data:

```
mayo<-read.table("mayonnaise.txt,"header=TRUE)
```

4. Alternatively you could read the data from a directory other than the working directory:

```
mayo<-read.table("C:\GUILLERMO\R_FILES\mayonnaise.
txt,"header=TRUE)
```

5. Open Source R-code: arrnon.R. (This R-function can be downloaded from the author's website.)
6. What is arrnon.R? It is a function that will allow estimating activation energy from data in a format such as shown in Table 7.3. The arrnon.R function has the following format and options:

```
arrnon <- function(data, Sensi=1,krefi=0.5,
ERi=10000,Tref=300,order="zero")
```

   - data: data with Time, Temperature, and Sensory columns
   - Sensi: initial value for Sensory descriptor
   - krefi: initial value for $k_{ref}$
   - ERi: initial value for $E_a/R$
   - Tref: chosen reference temperature in degrees Kelvin (within the accelerated storage range)
   - order: zero for zero-order kinetics; first for first-order kinetics

7. Use arrnon with default options for a zero-reaction rate:

```
ear<-arrnon(mayo)
```

8. To list the results type:

```
ear
```

|        | esti       | lower       | upper     |
| ------ | ---------- | ----------- | --------- |
| Sens0  | −1.1991637 | −9.7067061  | 7.31E+00  |
| Kref   | 0.2155167  | 0.1233301   | 3.08E-01  |
| ER     | 8726.65747 | 7339.58285  | 1.01E+04  |

For the present mayonnaise data the following equation results:

$$Ox = Ox_0 + k_{ref} \exp\left(-\frac{E_a}{R}\left(\frac{1}{T} - \frac{1}{T_{ref}}\right)\right) \cdot t$$

$$= -1.199 + 0.216 \exp\left(-8727\left(\frac{1}{T} - \frac{1}{300}\right)\right) \cdot t$$

(7.7)

The $E_a/R$ ± 95% confidence interval is 8727 ± 1387; and the $E_a$ ± 95% confidence interval is 17279 ± 2746 cal/mol. These values can be compared with those obtained using the simple linear regression method presented in Section 7.2.3: 13,470 ± 6710 cal/mol. On the one hand this last confidence band includes the non-linear estimate, and on the other hand the confidence interval obtained with the non-linear estimation is significantly narrower.

Suppose that it is of interest to know if the mayonnaise stored for 3 months at 25°C will develop an oxidized flavor value above the cut-off point, which was estimated to be 15 on the 0 to 60 scale. Equation 7.7 can be used to answer this question:

$$Ox = -1.199 + 0.216 \exp\left(-8727\left(\frac{1}{298} - \frac{1}{300}\right)\right) \cdot 90 = 14.8$$

Another angle to the above question is, what would the storage time be to reach the cut-off point at 18°C? In this case Equation 7.7 can also be used:

$$Storage\ time = \frac{15 + 1.199}{0.216 \exp\left(-8727\left(\frac{1}{291} - \frac{1}{300}\right)\right)} = 184\ days$$

Assuming first-order kinetics, the R-function arrnon is as follows:

```
ear<-arrnon(mayo,krefi=0.02,order="first")
```

For which the result after typing *ear* is

|        | esti       | lower      | upper      |
|--------|------------|------------|------------|
| lSens0 | 1.4271E+00 | 7.4586E-01 | 2.1084E+00 |
| Kref   | 1.3932E-02 | 6.6338E-03 | 2.1230E-02 |
| ER     | 8.2499E+03 | 6.5926E+03 | 9.9071E+03 |

Equation 7.6 for the mayonnaise data based on first-order kinetics is:

$$\ln Ox = 1.427 + 0.01393 \exp\left(-8250\left(\frac{1}{T} - \frac{1}{300}\right)\right) \cdot t$$

Suppose two temperatures are defined:

$T_a$ = temperature used to accelerate reaction rates
$T_n$ = normal storage temperature

The Arrhenius equation can be used to define the acceleration factor (AF), which is the ratio of the reaction rate constants at these two temperatures:

$$AF = \frac{k(T_a)}{k(T_n)} = \exp\left(\frac{E_a}{R}\left(\frac{1}{T_n} - \frac{1}{T_a}\right)\right) \tag{7.8}$$

Equation 7.4 can be expressed in terms of the acceleration factor as defined in Equation 7.8:

$$Ox_{SL} = Ox_0 + AF.k(T_n).SL(T_a)$$
$$Ox_{SL} = Ox_0 + k(T_n).SL(T_n) \tag{7.9}$$

where $Ox_{SL}$ is the oxidized flavor when the product reaches its shelf life, either at $T_a$ or $T_n$. It is assumed that the oxidized flavor at which a given consumer will reject the mayonnaise is the same regardless of the temperature at which it was stored; that is, the consumer tastes the mayonnaise and decides to reject it without knowing its temperature history. Thus, the left terms of Equation 7.9 can be assumed to be equal, and it thus follows that:

$$SL(T_n) = AF.SL(T_a) \tag{7.10}$$

The meaning of Equation 7.10 is that the shelf life at the normal storage temperature condition is equal to shelf life at the accelerated condition multiplied by the accelerating factor. Thus if the shelf life of a food product at an accelerated temperature and the activation energy are known, then the shelf life at a different temperature can be estimated.

## 7.3    The use of $Q_{10}$

$Q_{10}$ is defined as the ratio of the reaction rate constants at temperatures differing by 10°C, or the change in shelf life when the food is stored at a temperature differing by 10°C:

$$Q_{10} = \frac{k_{T+10°C}}{k_T} = \frac{shelf-life_T}{shelf-life_{T+10°C}} \qquad (7.11)$$

In the food industry $Q_{10}$ has been used as a more direct index of how the reaction rate or shelf life changes with temperature.

If the $k_{T+10°C}$ and $k_T$ values are replaced by the Arrhenius equation (Equation 7.2), $Q_{10}$ can be expressed as:

$$Q_{10} = \exp\left(\frac{E_a}{R} \times \frac{10}{T(T+10)}\right) \qquad (7.12)$$

This equation indicates that if $E_a$ is known, then $Q_{10}$ can be calculated, although the estimated value will depend on the temperature $T$ that is considered. Table 7.4 shows how $Q_{10}$ varies for different temperatures and $E_a$ values. Each particular situation will define whether these variations are of practical importance.

The $Q_{10}$ approach in essence introduces a temperature dependence equation of the form:

$$k = k_0 \exp(bT) \text{ or } \ln(k) = \ln(k_0) + bT \qquad (7.13)$$

where $k$ is the reaction rate constant at temperature $T$, and $k_0$ and $b$ are regression constants. Equation 7.13 implies that if $\ln(k)$ is plotted versus temperature (instead of $1/T$ of the Arrhenius equation) a straight line is obtained. Similarly, Labuza (1982) stated that if log(shelf life) is plotted

*Table 7.4* $Q_{10}$ at Different Temperatures and Different Activation Energy Values (Equation 7.12)

| $E_a$ (kcal/mol) | $Q_{10}$ at 5°C | $Q_{10}$ at 20°C | $Q_{10}$ at 40°C |
|---|---|---|---|
| 10 | 1.87 | 1.76 | 1.64 |
| 20 | 3.51 | 3.1 | 2.7 |
| 30 | 6.58 | 5.47 | 4.45 |

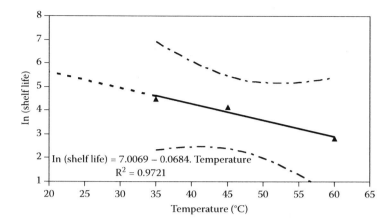

*Figure 7.4* ln(shelf life) versus storage temperature for bottled sunflower oil exposed to light. Solid line is the regression for experimental points, dashed line is the extrapolation to predict shelf life at 20°C, and curved lines are 95% confidence intervals.

versus storage temperature, a straight line is usually obtained. This is true; however, it was not made clear that this straight line is usually obtained with only three or, at the most, four experimental points corresponding to the tested temperatures. As seen above (Section 7.2.3), when a limited number of experimental points is used in accelerated test estimations, confidence intervals are very wide.

Ramírez et al. (2001) estimated shelf lives of bottled sunflower oil illuminated 12 hours a day to be 91 days, 60 days, and 17 days for storage at 35°C, 45°C, and 60°C, respectively. These values are plotted in Figure 7.4. The extrapolation shown in Figure 7.4 was used by Ramírez et al. (2001) to estimate a shelf life of 281 days at 20°C. However they did not report confidence intervals for this estimation. As observed in Figure 7.4, these confidence intervals are extremely wide. If they are calculated and then back transformed to the shelf-life scale of days, the confidence intervals result in 4 days for the lower bound and 19,000 days for the upper bound. Clearly this extrapolation is of little value.

Alternatively a nonlinear estimation can be performed using the following equation:

$$\text{Shelf life} = SL_0 \cdot \exp(aT)$$

where $T$ is the temperature and $SL_0$ and $a$ are the regression coefficients. The estimated shelf life at 20°C ± 95% confidence intervals is 215 ± 730 days. Here again the confidence intervals imply that the extrapolation is of little value.

In the above calculations three temperatures were used to estimate shelf life at a room temperature of 20°C. If three temperatures cannot be tested, at least two should be tested. If the sunflower oil experiment described by Ramírez et el. (2001) had been performed with only two temperatures, 45°C and 60°C, an extrapolation using ln(shelf life) versus temperature would have given an estimated shelf life of 491 days instead of the 281 days estimated with the three temperatures. The mayonnaise data presented in Table 7.1 arises from an experiment performed with three temperatures. In Section 7.2.3 the activation energy was calculated using a non-linear approach. If this approach is applied to only two of the three experimental temperatures, 35°C and 45°C, an activation energy value of 8,312 cal/mol is obtained instead of the value of 17,277 cal/mol obtained from the three temperatures. Using the data in Table 7.1 and a cut-off point of 15 on the oxidized flavor scale, the estimated shelf lives at 20°C and 35°C were 184 days and 23 days, respectively. Shelf life at 20°C can also be estimated using Equation 7.10 with an activation energy of 8,312 cal/mol (based on data from 35°C and 45°C only) and a shelf life of 23 days at 35°C:

$$SL(T_{20°C}) = AF.SL(T_{35°C}) = 2 \cdot 23 = 46 \text{ days}$$

This value is far different from the value of 184 days based on the 20°C kinetic data presented in Table 7.1. From these two real case examples it is clear that the use of only two temperatures in estimating shelf lives at a third temperature is likely to lead to wide over- or underestimates.

$Q_{10}$ is a means of communicating variations in shelf life in relation to temperature changes. If there are plans of exporting a food product to a country where ambient temperatures are likely to be 10°C higher than local temperatures, it is easier for company staff to understand that "shelf life will be reduced by a factor of 2.5 than for them to grasp the implications of an activation energy of 15 kcal/mol. However, using Equation 7.10 is just as easy to use as the $Q_{10}$ approach and is not restricted to 10°C intervals. Our recommendation is to perform activation energy calculations as described in Section 7.2.4, and if a $Q_{10}$ value is necessary for communication reasons, it can be calculated using Equation 7.12.

# 7.4  Survival analysis accelerated storage model

## 7.4.1  Accelerated storage model

Nelson (1990) and Meeker and Escobar (1998) have applied survival analysis to accelerated storage of industrial elements and materials such as electronic components, lamps, and automobile parts. Hough et al. (2006a) applied

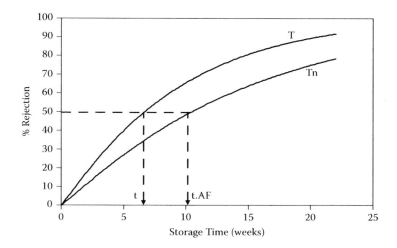

*Figure 7.5* Percent rejection versus storage time for a food product stored at temperatures $T$ and $T_n$. For a 50% rejection probability there is a corresponding storage time of $t$ for the $T$ condition and a storage time of $t \cdot AF$ (acceleration factor, see Equation 7.8) for the $T_n$ condition.

these models in estimating the shelf lives of foods at different temperatures based on the consumers' decision of accepting or rejecting the product.

The basic equation relating rejection probabilities at different temperatures is the following (Meeker and Escobar 1998):

$$F(t,T) = F(t.AF, T_n) \qquad (7.14)$$

Equation 7.14 means the rejection probability corresponding to time $t$ at the temperature condition $T$ is equal to the rejection probability corresponding to time $t$ multiplied by the acceleration factor ($AF$, Equation 7.8) at the normal storage temperature $T_n$. This is graphically shown in Figure 7.5. If $F(t)$ follows a log location-scale distribution (for example, the log-normal (Equation 4.2) or the Weibull (Equation 4.3) distributions) with parameters $\mu$ and $\sigma$, then Equation 7.14 can be expressed as:

$$\Phi\left(\frac{\ln(t) - \mu_T}{\sigma}\right) = \Phi\left(\frac{\ln(t.AF) - \mu_{T_n}}{\sigma}\right)$$
$$= \Phi\left(\frac{\ln(t) - (\mu_{T_n} - \ln(AF))}{\sigma}\right) \qquad (7.15)$$

If the log-normal distribution is considered, then $\Phi$ is the normal cumulative distribution; if the Weibull distribution is considered, then $\Phi$ is the $F_{sev}$ distribution.

From Equation 7.15 if $\mu_T$ is expressed as is customary for log location-scale regression models with inclusion of covariates (Meeker and Escobar 1998), that is:

$$\mu_T = \beta_0 + \beta_1.Z$$

and considering that from Equation 7.8:

$$\ln(AF) = \frac{E_a}{R}\left(\frac{1}{T_n} - \frac{1}{T}\right)$$

then:

$$\beta_0 = \mu_{T_U} - \frac{E_a}{R.T_n}$$

$$\beta_1 = \frac{E_a}{R} \tag{7.16}$$

$$Z = \frac{1}{T}$$

In Equation 7.15 it is assumed that $\sigma$ remains constant over different temperatures. Different $\sigma$ values at different temperatures would result in distribution lines with different slopes. Such lines would cross, resulting in lower probability of rejection for the higher temperature beyond the time where the lines cross. Such crossing is physically implausible (Nelson 1990). Thus, a constant value for $\sigma$ is usually assumed. A deviance test described below (Meeker and Escobar 1998) can be performed to test this assumption.

## 7.4.2 Experimental data

The model will be illustrated with data obtained from a storage study of minced meat. A homogeneous batch of minced beef was divided in portions of approximately 30 grams, which were placed in 100-ml clear glass bottles with screw caps. These bottles were frozen to –18°C. The bottles were removed from the freezer at different time intervals so that all samples with different storage times were ready at the same time. After thawing for 2 hours, the samples were placed at one of three storage temperatures for the following times:

2°C: 0, 24, 48, 96, 144, 192, and 240 hours
9°C: 0, 24, 48, 72, 96, 120, and 144 hours
19°C: 0, 6, 12, 18, 24, 36, and 48 hours

Sixty consumers who bought minced meat and/or prepared food using minced meat at least once every 2 weeks were recruited. Each consumer received the 21 bottles with minced meat (7 for each storage temperature) monadically in random order. For each sample they had to look at the bottle and answer the question: "Would you normally consume this product: Yes or No?" Their acceptance or rejection of the sample was based exclusively on appearance.

## 7.4.3   Calculations using R

In Section 5.3.3 the survival model for quantitative covariates was presented. For accelerated storage calculations the model and R-function are basically the same. The first step in performing calculations is to have the raw data in an Excel spreadsheet in a format as shown in Table 7.5. To obtain the data in this format the following procedure can be followed:

1. Obtain the raw data from consumers in the format shown in Table 7.6 using a separate table for each temperature.
2. Process the raw data for each group using the R function *sslife* (Section 4.8 and Figure 4.6), and store the results in, for example, res19 for the data corresponding to minced meat stored at 19°C.
3. Display the censored data by typing > res19$censdata.
4. Select the censored data with the mouse and copy and paste the data to an Excel spreadsheet.

*Table 7.5* Portion of the Data Used for Survival Analysis Accelerated Storage Calculations

| Consumer | tlow | thigh | cens | censcod | Kelvin | Celsius |
|---|---|---|---|---|---|---|
| 1 | 240 | 240 | right | 0 | 0.0036364 | 2 |
| 2 | 96 | 240 | interval | 3 | 0.0036364 | 2 |
| 3 | 96 | 144 | interval | 3 | 0.0036364 | 2 |
| ... | ... | ... | ... | ... | ... | ... |
| 1 | 48 | 72 | interval | 3 | 0.0035461 | 9 |
| 2 | 48 | 72 | interval | 3 | 0.0035461 | 9 |
| 3 | 72 | 144 | interval | 3 | 0.0035461 | 9 |
| ... | ... | ... | ... | ... | ... | ... |
| 52 | 6 | 6 | left | 2 | 0.0034247 | 19 |
| 53 | 6 | 24 | interval | 3 | 0.0034247 | 19 |
| 54 | 6 | 24 | interval | 3 | 0.0034247 | 19 |

*Note:* The full data can be downloaded from the editor's Web site: mince.xls.

*Table 7.6* Acceptance/Rejection Data for Minced Meat
Stored at 19°C for 6 of the 60 Consumers

| Consumer | t0 | t6 | t12 | t18 | t24 | t36 | t48 |
|----------|-----|-----|------|------|------|------|------|
| 1 | yes | yes | no | no | no | no | no |
| 2 | yes | yes | yes | yes | no | no | no |
| 3 | yes | yes | yes | yes | no | no | no |
| ... | ... | ... | ... | ... | ... | ... | ... |
| 52 | yes | no | no | no | no | no | no |
| 53 | yes | yes | no | yes | no | no | no |
| 54 | yes | yes | no | yes | no | no | no |

5. Using the *Data/Text in columns* utility, transform the pasted data to the corresponding columns.
6. Once the data corresponding to the three temperatures has been processed in this way, the censored data can be appended and the necessary columns shown in Table 7.5 can be created.

The instructions to analyze the data using the R-statistical package are the following:

1. Save the data table mince.xls (Table 7.5) from Excel to a tab-delimited text file (extension .txt).
2. Open R.
3. Change working directory if necessary.
4. Read raw data:

```
min<-read.table("mince.txt",header=TRUE)
```

Alternatively you could read the data from a directory other than the working directory:

```
min<-read.table("C:\GUILLERMO\R_FILES\mince.txt",
header=TRUE)
```

The raw data must have the following columns: consumer, lower time of the interval, upper time of the interval, type of censorship in words, type of censorship in code, and inverse of Kelvin units. In the case of mince.txt, there is an additional column indicating degrees Celsius. But this last column and other columns are not necessary.
5. Go to the File Menu and open a New Script. This will bring up an empty window.
6. Introduce the text shown in Figure 5.7 in the New Script window. Once it has been introduced, save this window as *sslcov.R* (Steps 5

and 6 can be omitted if sslcov.R has already been loaded into R as shown in Section 5.3.3.2).

7. Open Source R-code: sslcov.R.
8. What is sslcov.R?
   a. It is a procedure to deal with any linear single covariate. In the case of accelerated studies following the Arrhenius model, the linear covariate is the inverse of temperature in Kelvin units.
   b. sslcov <- function(daten,predict=1,model="weibull,"percent= c(10,25,50)).
   c. daten: censored data.
   d. predict: the value of the covariate for which I want to estimate shelf lives. It can be one of the experimental values or a new value.
   e. model: parametric model of choice (weibull, exponential, gauss-ian, logistic, log-normal or loglogistic).
   f. percent: percent rejection values for which I want estimated shelf lives. For example, if I want estimated shelf lives for per-cent rejections = 10%, 20%, 30%, 40%, and 50%, then percent = c(10,20,30,40,50). If I want a table with estimated shelf lives for a sequence of percent rejections from 10 to 50% at 1% increments, then percent = c(seq(10,50,by=1)).
9. For the mince.txt data, we shall use sslcov for estimating shelf lives at 12°C (1/285 = 0.00350877) using the log-normal distribution (it gives the lowest loglikelihood for the full model, very similar to the loglogistic):

```
resmin<-sslcov(min,predict= 0.00350877,model="
lognormal")
```

The results can be displayed by separately typing: resmin$musig, resmin$loglike, resmin$chiprob100, resmin$predict.for, and resmin$slives.

10. Or they can all be listed together:

```
resmin
```

$musig

|   | value    | beta0     | sigma     | beta1    |
|---|----------|-----------|-----------|----------|
| 1 | estimate | −23.59574 | 0.6999511 | 7719.728 |
| 2 | lower    | −28.37328 | 0.6098866 | 6368.553 |
| 3 | upper    | −18.8182  | 0.8033157 | 9070.902 |

*The estimate under beta1 is $E_a/R$.*

`$loglike`

`[1]  271.2019  228.5282`

These are the loglikelihood values corresponding to the reduced model without the covariate (higher loglike) and the full model including the covariate (lower loglike). This second value is used to choose the most appropriate model; the lower it is the better the fit.

`$chiprob100`

`[1]  0`

This is the percent significance level of the chi-square test comparing the full model including the covariate to the reduced model without the covariate. In this case, percent significance is very low and thus the full model is significant and we have to use the covariate.

`$predict.for`

`[1]  0.00350877`

`$slives`

| percent | estimate | lower ci | upper ci | serror |
|--------:|---------:|---------:|---------:|-------:|
| 10 | 13.38303 | 11.18065 | 16.01925 | 1.227711 |
| 25 | 20.46879 | 17.7509  | 23.60283 | 1.487801 |
| 50 | 32.81912 | 29.0406  | 37.08926 | 2.048119 |

These are the estimated shelf-lives corresponding to 12°C, that is, $1/285 = 0.00350877$.

The $E_a/R \pm 95\%$ confidence intervals was $7720 \pm 1351$; thus, considering $R = 1.98$, $E_a = 15{,}286 \pm 2676$ cal/mol. These last cal/mol units have no meaning in the context of sensory shelf life estimated from consumer data. The $E_a$ value is an indication of how consumers' acceptance or rejection for the appearance of raw minced meat changes with storage time as a function of storage temperature. $E_a$ can be used in estimating shelf life at temperatures other than those experimented. For example, the shelf life of the minced meat at 19°C was 17 hours, considering a rejection probability of 50%. If an estimation of the shelf life at 15°C was wanted, Equation 7.10 can be used:

$$SL(T_{15°C}) = AF.SL(T_{19°C}) = \exp\left(7720\left(\frac{1}{288} - \frac{1}{292}\right)\right) \times 17 = 24.5 \text{ h}$$

If it were of interest to plot percent rejection versus storage time for a temperature of 12°C, the value of μ to introduce in the log-normal model which was chosen for the minced meat data, can be calculated as follows (see Equations 7.15 and 7.16):

$$F(t, 285) = \Phi\left(\frac{\ln(t) - \mu_{285}}{\sigma}\right) = \Phi\left(\frac{\ln(t) - \left(\beta_0 + \dfrac{E_a}{R}\dfrac{1}{285}\right)}{\sigma}\right) =$$

$$= \Phi\left(\frac{\ln(t) - \left(-23.596 + \dfrac{7720}{285}\right)}{0.70}\right) = \Phi\left(\frac{\ln(t) - 3.492}{0.70}\right)$$

(7.17)

Figure 7.6 shows estimated percent rejection versus storage time for storage temperatures of 12°C and 19°C. For this last temperature $\mu_{292}$ was calculated as shown for 12°C (Equation 7.17) and its value was = 2.842.

One issue that can be of concern is whether the Arrhenius model actually fits the data adequately. Meeker and Escobar (1998) propose a significance test to compare individual fits at each temperature with the Arrhenius fit. Table 7.7 presents μ and σ values corresponding to the individual and Arrhenius fits to the minced meat data. The individual fits have no constraints. The Arrhenius model constrains μ to be a linear function of $1/T$ K and σ to be the same for all temperatures. In absolute values the total likelihood of the unconstrained model will always be lower than the log likelihood of the constrained model. If this difference is large, then

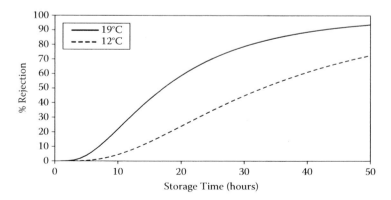

*Figure 7.6* Percent rejection versus storage time for minced meat stored at 12°C and 19°C.

*Table 7.7* Log-Normal Parameters μ and σ Corresponding to Individual Fits for Each Temperature and to the Arrhenius Fit to the Complete Data Obtained from Acceptance/Rejection of the Appearance of Minced Meat, and Loglikelihood Values for the Individual Fits

|  | Individual fits | | | Arrhenius fit[a] | |
|---|---|---|---|---|---|
| Temperatures | μ | σ | Log likelihood | μ[b] | σ |
| 2°C | 4.475 | 0.799 | 75.5 | 4.477 | 0.70 |
| 9°C | 3.786 | 0.606 | 72.8 | 3.780 | 0.70 |
| 19°C | 2.845 | 0.687 | 78.9 | 2.842 | 0.70 |

[a] The loglikelihood value corresponding to the Arrhenius fit was 228.5.
[b] These μ values were calculated as shown in Equation 7.17.

there is evidence of a lack of fit of the constrained Arrhenius model. This comparison can be made with an "omnibus" likelihood ration test. The likelihood of the unconstrained model is the sum of the individual likelihoods obtained from the individual fits at each temperature ($L_2$, $L_9$, and $L_{19}$), that is:

$$L_{unconstrained} = L_2 + L_9 + L_{19} = 75.5 + 72.8 + 78.9 = 227.2$$

The log likelihood of the constrained model was:

$$L_{constrained} = 228.5$$

The test statistic:

$$Q = 2(L_{constrained} - L_{unconstrained}) = 2.6$$

follows a chi-squared distribution with three degrees of freedom. These degrees of freedom come from the difference between the six parameters calculated for the unconstrained model (μ and σ values for each temperature) and the three parameters calculated for the constrained model ($\beta_0$, $E_a/R$, and σ). The probability corresponding to this value of chi-square is 0.46, thus there is no evidence of inadequacy of the constrained model, relative to the unconstrained model.

Another issue of interest is whether two temperatures would be sufficient instead of the three temperatures used in the minced meat study. Activation energy estimated from the three temperatures was $E_a = 15,286 \pm 2676$ cal/mol. If only two temperatures were used to estimate activation energy, for example, 9°C and 19°C, the resulting activation energy would have been $E_a = 15,192 \pm 4322$ cal/mol. The mean values were similar for two or three temperatures, but the confidence interval for the two-temperature estimation was substantially higher. Table 7.8 shows how using two

*Table 7.8* Shelf-Life Values for the Appearance of Minced Meat Estimated at 5°C for 50% Consumer Rejection Using Two or Three Prediction Temperatures and Their 95% Confidence Intervals

| Number of prediction temperatures | Shelf life estimation (h) | Lower confidence limit (h) | Upper confidence limit (h) |
|:---:|:---:|:---:|:---:|
| 2 | 65 | 49 | 86 |
| 3 | 65 | 56 | 75 |

or three temperatures influenced shelf-life estimations at 5°C considering a 50% rejection probability. As with the activation energy, the confidence interval for the two-temperature estimation was substantially higher. Three temperatures give sufficiently narrow confidence intervals, while two temperatures can give confidence intervals that are too wide.

An important aspect of survival analysis methodology applied to accelerated shelf-life models is that experimental sensory work is relatively simple. In the case of minced meat 60 consumers evaluated samples stored for seven different times at each temperature, answering yes or no to whether they would consume the samples. This information was sufficient to model the probability of consumers rejecting the products with different storage times at different temperatures. Shelf-life and activation energy estimations with their corresponding confidence intervals were obtained without having to count on a trained sensory panel. Data from a trained panel can be of great value to understand the mechanisms underlying consumer's acceptance or rejection of products stored at different temperatures.

## 7.5   Potential pitfalls of accelerated shelf-life testing

Most published work on accelerated shelf-life testing highlights the risks involved in such testing. The ASTM E2454 Standard (2005, p. 5) indicates that these tests are only approximations of how a product may behave under normal storage conditions and also recommends that before determining shelf life based on accelerated conditions, to "establish the sensory, chemical, and mathematical relationships between accelerated conditions and typical storage conditions to ensure a high degree of reliability and validity in predicting shelf life." In Section 3.1 the complexity of food products and their modes of deterioration were discussed. Their main characteristics are (a) the foods are heterogeneous, (b) the chemical reactions involved in their deterioration have complex kinetics, (c) rheological and textural properties are not simply modeled, (d) biological changes continue during food storage, and (e) the deterioration of a food during storage is often the result of a number of simultaneous processes.

Thus, there are factors relevant to food and food quality loss reactions that can cause significant deviations from an Arrhenius behavior with temperature (Taoukis et al. 1997). The IFST (1993) guidelines have stated that there are practical problems, both general and product-specific, that have limited the potential applications and usefulness of accelerated shelf-life determination.

Meeker and Escobar (1998) present an interesting summary of potential pitfalls of accelerated studies that could cause an accelerated test to lead to seriously incorrect results.

## 7.5.1   Pitfall 1: Multiple deterioration modes

High levels of accelerating variables like temperature can induce changes in a food product that would not be observed at normal storage conditions. In some cases different modes occur due to a fundamental change in the way the food components deteriorate at high levels of the accelerating variable. For example, instead of simply accelerating a chemical process that leads to the presence of a sensory defect, increased temperature may lead to changes in the material properties such as a change of phase. This would be the case of a dairy product placed at temperatures above 37°C. At this temperature milk fat starts melting and chemical reactions such as lipid oxidation evolve in a different manner. At the other extreme, on freezing, reactants are concentrated in the unfrozen liquid resulting in a higher rate of reaction (e.g., fat oxidation in frozen meat) at a lower (in this case, subzero) temperature.

If other rejection modes are caused at high levels of an accelerating variable, they can be accounted for in the data analysis by treating the new rejection mode as a censored observation. For example, in a shelf-life study of bottled vegetable oil a discoloration of the label can occur due to a high storage temperature. If the study is designed for consumers to take the stored product home, they would probably reject the product regardless of the oxidized flavor that is the target of the study. If this were detected, label data would be right-censored and some measure like a change of labels would have to be taken. If these other rejection modes are present but not recognized in the data analysis, seriously incorrect conclusions are likely.

## 7.5.2   Pitfall 2: Failure in quantifying uncertainty

Basing decisions on point estimates alone with no statistical confidence bounds can be seriously misleading. Section 7.1.1 has the title "Acceleration Factor Fallacy" related to food companies that use, or aspire to use, a single acceleration test condition. When this factor is applied, it is not accompanied by any uncertainty calculations; usually the greatest uncertainty lies in where the factor originated. In Section 7.3 uncertainty calculations were introduced to the empirical model of ln(shelf-life) versus temperature. Confidence intervals of extrapolations beyond the experimented temperatures were so wide that estimations became meaningless. Unfortunately this failure in reporting uncertainty is common (Almeida-Domínguez et al. 1992; Sithole et al. 2005).

### 7.5.3 Pitfall 3: Degradation and rejection affected by unforeseen variables

A serious pitfall of accelerated testing is to assume a simple relationship between shelf life and the accelerating variables when the actual relationship is very complicated. For food products there will be a distribution of product-use conditions that can substantially influence consumers' rejection of the stored product. For example, it took one company several months to realize that the unusual number of shelf-life complaints received in a certain region of the country was due to the bad conditions of the road taken by the distribution truck that affected the texture and syneresis of their product. In Argentina there is a wide range of average temperatures between different parts of the country; thus, some products are submitted to a temperature stress. Most sensory shelf-life studies are done with coded samples, that is, with no brand available to consumers. If a prestigious brand with a sensory storage defect is presented to a consumer, will she assimilate this defect and accept the product or will she reject the product due to the high expectation of the brand? These and other examples illustrate that unforeseen variables can affect predicted shelf life.

### 7.5.4 Pitfall 4: Masked rejection mode

Figure 7.7 shows the possible results from a study on the rates of plastic and oxidized flavor development versus storage temperature in a product that could be mayonnaise. At the $T_{high}$ accelerated condition the rate of plastic flavor development is higher than the rate of oxidized flavor development. Due to the different activation energies, evidenced by the different slopes observed in Figure 7.7, at the $T_{low}$ room temperature the situation is reversed and the rate of oxidized flavor development is higher than the rate for plastic flavor. Thus it is clear that the true critical descriptor at room temperature was masked at the accelerated condition. In this case the food company would be receiving complaints related to oxidized flavor (very probably referred to as "rancid" by the consumers) before the established shelf life based on plastic flavor had been reached.

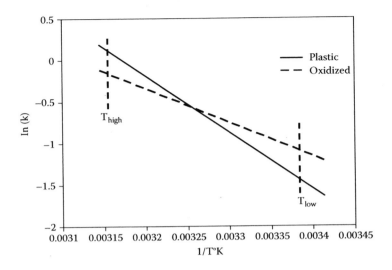

*Figure 7.7* Possible results from a study on the rates of plastic and oxidized flavor development versus storage temperature.

Food products in which these masking effects can occur are those that contain active microorganisms. For example, pasteurized milk has a variety of active microorganisms: psychrophiles, mesophiles, and thermophiles, which grow better at low, medium, and high temperatures, respectively. If milk is stored at 35°C with the intention of accelerating spoilage, mesophile and thermophile microorganisms will grow, generating specific off-flavors different from those generated during refrigerated storage, where psychrophile microorganisms grow faster. This change of the dominating microorganism with storage temperature could lead to erroneous predictions.

## 7.5.5   Pitfall 5: Comparisons that do not hold

In the food oil and fat industry an instrument (the Rancimat, by Metrohm AG, in Herisau, Switzerland) accelerates oxidation by exposing the sample to elevated temperatures while pumping air into it. Acceleration temperatures are in the range of 100°C to 120°C. In many cases this instrument is used for comparing alternatives, as would be the case of comparing the application of two antioxidants in a margarine formulation. The basic claim is that an accelerated test at 110°C and air circulation cannot pretend to approximate actual storage conditions of a solid pack at 10°C; but if Antioxidant A is better than Antioxidant B at the highly accelerated conditions, then the same would be true at 10°C. This situation is shown in Figure 7.8, where a lower rate of oxidation is shown for Antioxidant A,

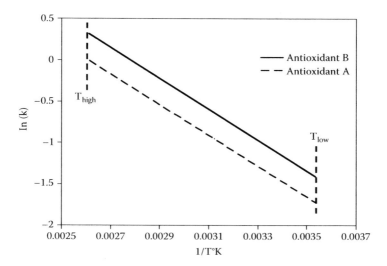

*Figure 7.8* Possible results from a study on the rates of oxidation versus storage temperature for two types of antioxidants that behave similarly at high and low temperatures.

both at the high and low temperatures. However, at elevated temperatures events occur that do not occur at 10°C, for example, the decomposition of some antioxidants (Arnaud et al. 2009). If this were the case for Antioxidant B, the situation could be like the one shown in Figure 7.9. In this case, if conclusions were based solely on the highly accelerated condition, Antioxidant A would be chosen mistakenly.

## 7.5.6   Pitfall 6: Increasing temperature can cause deceleration

One known product in which this occurs is bread (Taoukis et al. 1997). Retrogradation of the amylopectin and a redistribution of moisture between starch and gluten have been implicated in staling. Staling shows a negative temperature effect between 4°C and 40°C, achieving maximum rate at 4°C. Thus, pretending to accelerate bread staling by increasing storage temperature would have the inverse effect. In fact, negative activation energies have been reported for bread (Labuza 1982).

Another case where a temperature increase can cause deceleration is for food products that contain active enzymes and/or microorganisms. In Argentina and other countries food legislation states that yogurt should have living microorganisms. If an unwary experimenter were to store yogurt at a temperature ≥45°C with the idea that this would accelerate acid flavor development, he would find that no acceleration occurred due to the inactivation of responsible microorganisms above this temperature.

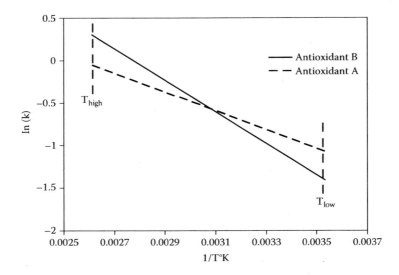

***Figure 7.9*** Possible results from a study on the rates of oxidation versus storage temperature for two types of antioxidants that behave differently at high and low temperatures.

These cases are not to be expected in practice as enzyme and microorganism inactivation temperatures are generally well known.

## 7.5.7   Pitfall 7: Drawing conclusions from pilot-plant samples

In the development of a food product shelf-life studies are usually initiated before the product is taken to full-scale production. Suppose a biscuit is to be launched on the market in a 2-month period. The company's experience indicates that the biscuit's shelf life will most probably be greater than 6 months. Thus an accelerated shelf-life test is called for with available samples coming from pilot-plant production. A number of factors can lead to seriously incorrect conclusions based on these pilot-plant samples:

* Cleanliness and care in producing pilot-plant samples versus production units may differ substantially.
* Ingredients used to manufacture the pilot-plant samples can later be changed for a number of reasons like cost or availability.
* When ingredient suppliers are called in to participate in the development of the new product, they may take special care in the high quality of their supply, and then under regular production these high quality standards may be slackened.

- Personnel closely linked to the development of the new product might have a tendency to discard pilot-plant samples they mistakenly consider to be non-representative of the quality they hope the product will have.

## 7.6 Conclusion on accelerated testing

At the Sixth Pangborn Sensory Science Symposium held in Harrogate, United Kingdom, in 2005, I coordinated a workshop on sensory shelf life (Hough et al. 2006b). One of the topics discussed at this workshop was accelerated testing. The conclusions were as follows:

1. Accelerated shelf-life tests should only be carried out in very simple systems and where the mechanisms involved in the acceleration test are well known.
2. In many cases pretending to carry out accelerated tests is tantamount to pretending to possess a time machine.
3. If accelerated testing worked in complex food products, you could make fine aged wine in an oven; this is not possible.
4. Participants expressed that retailers were becoming more flexible about the difficulties of date labeling, thus accepting that accelerated tests might not be adequate in many cases.

## References

Almeida-Domínguez, N.G., I. Higuera-Ciapara, F.M. Goycoolea, and M.E. Valencia. 1992. Package, temperature and TBHQ effects on oxidative deterioration of corn-based snacks. *Journal of Food Science* 57: 112–117.

Arnaud J.N., H. Lechat, and F. Lacoste. 2009. Evaluation of antioxidant efficiency in animal fat using a modified Rancimat test method. http://prestations.iterg.com/IMG/pdf/Poster_AOCS_Rancimat.pdf. Accessed July 9, 2009.

ASTM E2454 Standard. 2005. *Standard guide for sensory evaluation methods to determine the sensory shelf life of consumer products.* West Conshohocken, PA: American Society for Testing and Materials.

Fu, B., and T.P. Labuza. 1993. Shelf-life prediction: Theory and application. *Food Control* 4: 125–133.

Hough, G., L. Garitta, and G. Gómez. 2006a. Sensory shelf-life predictions by survival analysis accelerated storage models. *Food Quality and Preference* 17: 468–473.

Hough, G., D. van Hout, and D. Kilcast. 2006b. Workshop summary: Sensory shelf-life testing. *Food Quality and Preference* 17: 640–645

IFST. 1993. *Shelf life of foods: Guidelines for its determination and prediction.* London: Institute of Food Science & Technology.

Labuza, T.P. 1982. *Shelf-life dating of foods.* Westport: Food & Nutrition Press.

Martínez, C., A. Mucci, M.J. Santa Cruz, G. Hough, and R. Sánchez. 1998. Influence of temperature, fat content and commercial mayonnaise package material on the sensory shelf life of commercial mayonnaise. *Journal of Sensory Studies* 13: 331–346.

Meeker, W.Q., and L.A. Escobar. 1998. Statistical methods for reliability data. New York: John Wiley & Sons.

Mizrahi, S., and M. Karel. 1977a. Accelerated stability tests of moisture-sensitive products in permeable packages by programming rate of moisture content increase. *Journal of Food Science* 42: 958–963.

Mizrahi, S., and M. Karel. 1977b. Accelerated stability tests of moisture-sensitive products in permeable packages at high rates of moisture gain and elevated temperatures. *Journal of Food Science* 42: 1575–1579, 1589.

Nelson, W. 1990. *Accelerated testing: Statistical models, test plans and data analyses.* New York: John Wiley & Sons.

Ramírez, G., G. Hough, and A. Contarini. 2001. Influence of temperature and light exposure on sensory shelf life of a commercial sunflower oil. *Journal of Food Quality* 24: 195–204.

Sithole, R., M.R. McDaniel, and L. Meunier Goddik. 2005. Rate of Maillard browning in sweet whey powder. *Journal of Dairy Science* 88: 1636–1645.

Stone, H., and J.L. Sidel. 2004. *Sensory evaluation practices, 3rd edition.* San Diego: Elsevier Academic Press.

Taoukis, P., T.P. Labuza, and I.S. Saguy. 1997. Kinetics of food deterioration and shelf-life prediction. Chapter 10 in *Handbook of food engineering practice*, ed. K. Valentas, E. Rotstein, and R.P. Singh. Boca Raton, FL: CRC Press.

Taub, I.A., and R.P. Singh (eds). 1998. *Food storage stability.* Boca Raton, FL: CRC Press.

Yang, T.C.S. 1998. Ambient storage. Chapter 17 in *Food storage stability*, ed. I.A. Taub and R.P. Singh. Boca Raton, FL: CRC Press.

## chapter 8

# Other applications of survival analysis in food quality

## 8.1  Consumer tolerance limits to a sensory defect

Hough et al. (2004) presented the application of survival analysis statistics to study consumer tolerance limits to sensory defects. An important step in implementing a sensory quality control system is to define sensory standards or tolerance limits for the product (Lawless and Heyman 1998). A number of authors have emphasized the importance of establishing sensory specifications through consumer input (Costell 2002; Muñoz 2002; Weller and Stanton 2002).

One method for obtaining sensory specifications is to measure the cut-off point as described in detail in Chapter 6. Consumers are presented with samples covering a range of a specific sensory attribute, and they measure the acceptability of these samples by scoring them on a hedonic scale. These scores are then correlated versus intensity measurements of the same samples given by a trained sensory panel. Alternatively, the trained panel measurements can be replaced by a chemical or physical index of the samples. The specification is obtained by calculating the cut-off point (Section 6.4) or by setting a minimum level of acceptability on the chosen hedonic scale (Muñoz et al. 1992). This methodology is illustrated in Figure 8.1 for the sensory specification of dark color in UHT milk.

In the routine consumption of a food product, consumers do not measure acceptability on a scale. They do not eat a piece of cheese and say, "This cheese is a 6 on my hedonic scale of 1 to 10; therefore, I won't buy this brand again." Rather, their judgments constitute acceptance or rejection of the product. Thus the question is, how high can the intensity or concentration of a sensory defect be before a consumer rejects the product? Survival analysis statistics can be used to answer this question.

### 8.1.1  Survival analysis model

In Chapter 4 survival analysis was presented for modeling consumer rejection probability as a function of the storage times of food products. In these cases the outcome variable of interest was time until the consumer

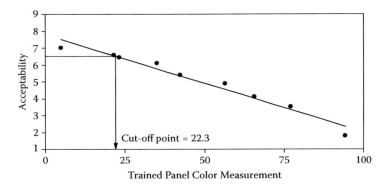

*Figure 8.1* Sensory acceptability of UHT milk versus dark color measured by a trained sensory panel.

rejection event occurred. Time can be replaced by other variables in the system under study, for example, distance to failure in vehicle shock absorbers (Meeker and Escobar 1998) or, as analyzed by Hough et al. (2004), concentration of a sensory defect. In sensory shelf life (SSL) the rejection function was the probability of a consumer's rejecting a food product stored for less than time $t$. If $C$ is the concentration of sensory defect at which a consumer rejects the sample, the failure function $F(c)$ can be defined as the probability of a consumer (or proportion of consumers) rejecting a food with the level of a sensory defect $<c$, that is $F(c) = P(C < c)$.

In Section 4.2 censoring in SSL studies was discussed. In a study to determine the concentration limit of a sensory defect, samples with different concentrations are presented to consumers. For example, concentrations could be 0, A, B, and C. If a consumer accepts the sample with concentration = A and rejects it with concentration = B, the exact concentration of rejection could be any value between A and B. This is defined as interval-censoring. If a consumer rejects the sample with concentration = A, rejection concentration is ≤A, and this is left-censoring. If the consumer accepts all concentrations, rejection would occur for a concentration >C, and the consumer's data is right-censored.

The model when concentration is the variable of interest is basically the same as the model described in Section 4.7, except that time is replaced by concentration. If the log-normal distribution is chosen for $C$, the rejection function is given by:

$$F(c) = \Phi\left(\frac{\ln(c) - \mu}{\sigma}\right) \tag{8.1}$$

where $\Phi(\cdot)$ is the standard normal cumulative distribution function, and $\mu$ and $\sigma$ are the model's parameters.

If the Weibull distribution is chosen, the rejection function is given by:

$$F(c) = F_{sev}\left(\frac{\ln(c)-\mu}{\sigma}\right)$$

where $F_{sev}(\cdot)$ is the rejection function of the extreme value distribution:

$$F_{sev}(w) = 1 - \exp(-\exp(w))$$

Thus, the rejection function for the Weibull distribution can be expressed as:

$$F(c) = 1 - \exp\left[-\exp\left(\frac{\ln(c)-\mu}{\sigma}\right)\right] \tag{8.2}$$

where $\mu$ and $\sigma$ are the model's parameters.

The parameters of the loglinear model are obtained by maximizing the likelihood function (Equation 4.1)—a mathematical expression that describes the joint probability of obtaining the data on the subjects in the study as a function of the unknown parameters of the model considered. To estimate $\mu$ and $\sigma$ for the log-normal or the Weibull distribution, the likelihood function is maximized by substituting $F(c)$ in Equation 4.1 by the expressions in Equations 8.1 or 8.2, respectively. For more details on likelihood functions see Klein and Moeschberger (1997) or Meeker and Escobar (1998).

## 8.1.2    Experimental data used to illustrate the methodology

To illustrate the methodology to be applied to estimate concentration limits using survival analysis statistics, data from UHT milk will be used. A commercial whole-fat UHT milk was provided by a local manufacturer in 1-liter cartons, all from the same batch. Hough et al. (2004) studied the following defects that can be present in UHT milk, either due to processing or storage problems: acid, caramel, cooked, dark color, lipolytic, and oxidized. Of these, caramel and cooked will be considered. Stock samples for these defects were prepared as follows:

- Caramel: 8 grams of Givaudan Roure (Munro, Argentina) caramel essence code 73865-33 per liter of UHT milk
- Cooked: UHT milk heated 15 minutes in a boiling water bath

These stock samples were diluted with UHT milk to obtain a series of nine concentrations for each defect: 0, 6, 9, 13, 20, 30, 44, 67, and 100% stock sample in UHT milk.

For each defect the nine concentrations were tested by a 50-member consumer panel. All consumers had drunk milk at least once in the last week, and were between 18 and 25 years old. The consumers were presented with the nine samples monadically in random order. Thirty ml of each sample was presented in a 70-ml plastic glass. For each sample they had to taste it and answer the question: "Would you normally consume this product: Yes or No?" It was explained that this meant that if they bought the product to drink it or if it was served to them at their homes, would they would consume it or not?

### 8.1.3   Rejection probability calculations

Table 8.1 presents data for 6 of the 50 consumers who tasted the caramel flavored UHT milks. The format of the raw data is very similar to survival analysis SSL data (see Table 4.1) and it can be processed using the sslife R-function (http://www.r-project.org/, accessed May 26, 2009) presented in Section 4.8 (Figure 4.6). Once the raw data is saved as a tab-delimited text file and read into R:

```
> car<-read.table("caramel.txt",header=T),
```

the sslife function can be written as:

```
> rescar<-sslife(car, tiempos=c(0,6,9,13,20,30,44,67,100),
codiresp=c("accept","reject"), model="lognormal")
> rescar
$censdata
```

*Table 8.1* Consumer Acceptance/Rejection Data for Caramel-Flavored UHT Milk

| consumer | c0 | c6 | c9 | c13 | c20 | c30 | c44 | c67 | c100 |
|---|---|---|---|---|---|---|---|---|---|
| 1 | accept | accept | reject | reject | accept | accept | accept | accept | accept |
| 2 | accept | accept | accept | reject | reject | accept | reject | reject | reject |
| 3 | reject | reject | accept | reject | reject | reject | reject | reject | reject |
| ... | ... | ... | ... | ... | ... | ... | ... | ... | ... |
| 48 | reject | accept | accept | accept | reject | reject | accept | accept | accept |
| 49 | accept | accept | reject | accept | accept | accept | accept | accept | reject |
| 50 | accept | reject | accept | accept | reject | reject | accept | accept | reject |

*Note:*   The full data can be downloaded from the editor's Web site: caramel.xls.

|   | id | ti | ts | cens | censcod |
|---|----|----|----|------|---------|
| 1 | 1 | 100 | 100 | right | 0 |
| 2 | 2 | 9 | 44 | interval | 3 |
| 3 | 4 | 44 | 67 | interval | 3 |
| ... | ... | ... | ... | ... | ... |
| 37 | 47 | 13 | 44 | interval | 3 |
| 38 | 49 | 6 | 100 | interval | 3 |
| 39 | 50 | 100 | 100 | left | 2 |

The type of censoring is similar to that obtained from SSL data, that is, left-, interval-, and right-censored consumers (see Section 4.8). The data from 11 of the 50 consumers was not considered because they rejected the sample with 0% caramel flavoring.

$musig

|   | value | mu | sigma |
|---|-------|-----|-------|
| 1 | estimate | 3.574694 | 0.9337544 |
| 2 | lower | 3.241587 | 0.6901572 |
| 3 | upper | 3.9078 | 1.2633314 |

These are the $\mu$ and $\sigma$ values with their 95% lower and upper confidence intervals corresponding to the log-normal distribution described by Equation 8.1.

$loglike
[1] 49.98814

The log-normal distribution had the lowest loglikelihood value; thus, it was chosen to model the data.

$slives

| percent | estimate | lower ci | upper ci | serror |
|---------|----------|----------|----------|--------|
| 10 | 10.78374 | 6.432577 | 18.07813 | 2.842637 |
| 25 | 19.00849 | 12.734923 | 28.37259 | 3.8845 |
| 50 | 35.68369 | 25.574283 | 49.7893 | 6.064522 |

In the first column of the previous table are the percent rejection values, in this case 10%, 25%, and 50%. In the second column are the estimated concentration values corresponding to each percent rejection. In the third and

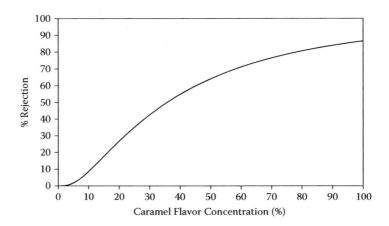

***Figure 8.2*** Percent rejection versus caramel flavor concentration in UHT milk.

fourth columns are the 95% lower and upper confidence limits, respectively. In the fifth column are the standard errors of the estimations.

The $\mu$ and $\sigma$ values calculated above can be used to plot percent rejection versus concentration as shown in Figure 8.2. To establish the concentration limit for caramel flavor a percent rejection value has to be chosen. In Section 4.9 values of 25% and 50% rejection probability were considered adequate for SSL determinations. This means that if a consumer tastes a product with a storage time corresponding to 50% rejection probability, there is a 50% probability that the consumers will reject the product. As previously discussed, this might appear to be too risky, but it must be pointed out that it refers to a consumer who tastes the product at the end of its shelf life. Distribution times usually guarantee that the proportion of consumers who taste the product close to the end of its SSL is small. Of this small proportion of consumers, 50% will reject the product and 50% will accept it. Concentration limits of a sensory defect are generally used as a quality control specification. If this is the case a 50% rejection probability is not acceptable. From quality control the product leaves the manufacturing plant and goes out to the consumer; it would not be desirable to have 50% of consumers rejecting the product. Thus it seems reasonable to adopt a 10% rejection probability for these cases, or even lower if it is an extremely critical product. For the present case of caramel flavor, a 10% rejection probability corresponds to a concentration limit of 11%, with 6% and 18% lower and upper 95% confidence intervals, respectively.

The above calculations and discussion concerned caramel flavor. What happened with the data from cooked flavor? (The full data can be downloaded from the editor's Web site: cooked.xls.) Calculations were performed as described above for the data from caramel flavor. The Weibull model had the lowest loglikelihood; results from this model were as follows:

|   | value | mu | sigma |
|---|-------|-----|-------|
| 1 | estimate | 5.35441 | 0.4999852 |
| 2 | lower | 4.324719 | 0.1443038 |
| 3 | upper | 6.384101 | 1.7323536 |

| percent | estimate | lower ci | upper ci | serror |
|---------|----------|----------|----------|--------|
| 10 | 68.66638 | 37.60681 | 125.3782 | 21.093 |
| 25 | 113.46324 | 73.30405 | 175.6234 | 25.28974 |
| 50 | 176.11886 | 77.20673 | 401.7506 | 74.10222 |

What can be observed from these results is that confidence intervals are particularly wide, both for the $\mu$ and $\sigma$ values and for the concentration limits corresponding to different percent rejections. For example, for a 10% rejection the concentration limit was estimated to be 69% with lower and upper 95% confidence limits of 37% and 125%, respectively. Going back to the raw data, once the data from consumers who rejected the 0% cooked flavor concentration were eliminated, 80% of the remaining consumers were right-censored. This introduces a high degree of uncertainty in the concentration at which a consumer rejects cooked flavor in UHT milk and this explains the wide range of the confidence limits reported above. The cooked flavor stock sample, corresponding to the 100% concentration, was prepared by heating UHT milk for 15 minutes in a boiling water bath. No commercial sample of UHT milk will reach this intensity of cooked flavor; thus, it was concluded that if 80% of the consumers found this sample acceptable (corresponding to right-censored observations), then cooked flavor was not critical for UHT consumers.

Hough et al. (2004) reported that for the oxidized defect, 41% of the consumers were right-censored; that is, a large proportion of consumers found the maximum concentration of this defect acceptable or below their threshold. Due to the high proportion of right-censored data, confidence limits, as was the case for the cooked flavor, were wide. What occurred for this defect was that the concentration range chosen was too low, probably below the threshold for many consumers; the stock sample should have had a higher oxidized flavor intensity. This eventuality can occur in SSL studies if the final storage time is too short. If this was the case, a large proportion of consumers would find the final sample acceptable, resulting in a high proportion of right-censored data.

### 8.1.4   Conclusions

An important aspect of this methodology is that experimental sensory work is relatively simple as no trained sensory panel work is necessary. In the present UHT milk example (Hough et al. 2004) 50 consumers were recruited for each defect. As seen in Section 5.1, the recommended number of consumers should have been approximately 120. If this number of consumers had been used, confidence limits of the estimated concentrations limits would have been narrower. The choice of the concentration range used to estimate the limits needs special attention. If the concentration range is too low (as occurred for the oxidized flavor) a large proportion of consumers will present right-censored data and estimations will lack precision. A large proportion of right-censored data also occurs when the defect does not produce consumer rejection within reasonable concentration limits, as occurred for the cooked flavor.

## 8.2   Optimum concentration of ingredients in food products

Just-about-right (JAR) scales are often used in research and development of food products (Lawless and Heymann 1998; Rothman and Parker 2009). One of their uses is determining the optimum level of an ingredient. In this case the researcher prepares a series of samples with increasing levels of the ingredient, and these are presented to consumers who judge their level using a JAR scale.

Conner et al. (1986) used a non-structured line scale anchored with "So far above/below the right sweetness that I'd never drink it" at the extremes and "At this sweetness I'd always drink it" in the middle. Ratings for each consumer were correlated with sugar concentration by simple linear regression. If a consumer rates different concentrations with the same score, for example, in the middle of the scale, he is somehow penalized as they obtain a low correlation coefficient. Yet this consumer is expressing the fact that there is a range of appropriate concentrations that deserve the same score, either because he cannot discriminate among the different levels of the ingredient or because he does not really care about the difference. Hernandez and Lawless (1999) used a nine-point hedonic scale to determine preferred levels of capsaicin in different foods by correlating mean hedonic ratings with concentration. The optimum concentration was estimated from the maximum hedonic score. One drawback of this method is that consumers do not decide that the score of a product is 8 on a 1 to 10 acceptability score, and therefore they consider the level of the ingredient to be JAR. In this paper consumers were segmented as dislikers (hedonic ratings decreased monotonically with increasing concentration), likers (hedonic ratings increased monotonically with increasing concentration), and non-monotonic (exhibited inverted U type behavior). Using survival analysis concepts (Hough et al. 2004) all consumers can be considered part of the same population. Ideal concentration for dislikers is below the range used in the study, and their data are left-censored. Similarly, ideal concentration for likers is beyond the range used in the study, and their data are right-censored.

In the previous section survival analysis concepts were applied to sensory defects. For example, for the caramel flavor defect in UHT milk, consumers either accepted the sample or rejected it. Caramel flavor was considered a defect and the rejection was for a single reason: high caramel flavor. But, for a desirable attribute such as red in strawberry yogurt the consumer can find the color too light, JAR, or too dark. The rejection of the product can be for two reasons: too light or too dark.

For this last case Garitta et al. (2006) presented a model based on survival analysis statistics to determine the optimum concentration of a food ingredient. In optimizing the color intensity of a yogurt, there will not be an ideal color that will be considered acceptable by all consumers. The objective will be to find an optimum color that maximizes the proportion of consumers who find the color appropriate. At this optimum, one group of consumers will find it appropriate, another group will find it too light, and yet another will find it too dark.

## 8.2.1   Survival analysis model

In the examples presented in this book there was only one event of interest, either failure time in SSL (Chapters 4, 5, and 7) or concentration limits for sensory defects (Section 8.1). As mentioned above, when a consumer opens a pot or bottle of yogurt she can find the yogurt too light, JAR, or too dark. Thus there are two events of interest: the transition from too light to JAR, and the transition from JAR to too dark. The order we have presented these two events is arbitrary, supposing that concentration increases. If we suppose the concentration decreases the two events would be too dark to JAR and JAR to too light.

Food color measurements are usually performed with colorimeters and expressed in a tristimulus CIELAB color space: $L^*$, $a^*$, and $b^*$. The $L^*$ represents luminosity; yellow to blue components of a color are represented by $b^*$; and green to red components of a color are represented by $a^*$ (Hutchings 1994). In the present work, we will assume that a yogurt's red color is represented by the parameter $a^*$, in positive values.

Let $A$ be the random variable representing the optimum color value for a consumer. Assume that $A$ is absolutely continuous with distribution function $F$. For each value of color $a^*$, there will be two rejection functions:

$R_l(a^*)$ = probability of a consumer (or proportion of consumers) rejecting a yogurt with a color = $a^*$ because it is too light; that is $R_l(a^*)$ = $P(A > a^*) = 1 - F(a^*)$

$R_d(a^*)$ = probability of a consumer (or proportion of consumers) rejecting a yogurt with a color = $a^*$ because it is too dark; that is $R_d(a^*)$ = $P(A < a^*) = F(a^*)$.

In consumer language, "too light" would mean low intensity of red color and "too dark," high intensity of red color.

With two rejection functions there will be two likelihood functions: $L_l$ (light colors) and $L_d$ (dark colors).

$$L_l = \prod_{ie\,R}\left(R_l(r_i)\right)\prod_{ie\,L}\left(1-R_l(l_i)\right)\prod_{ie\,I}\left(R_l(l_i)-R_l(r_i)\right) \tag{8.3a}$$

$$L_d = \prod_{ie\,R}\left(1-R_d(r_i)\right)\prod_{ie\,L}R_d(l_i)\prod_{ie\,I}\left(R_d(r_i)-R_d(l_i)\right) \tag{8.3b}$$

In both Equation 8.3a and Equation 8.3b, $R$ is the set of right-censored observations, $L$ is the set of left-censored observations, and $I$ is the set of interval-censored observations.

If the normal distribution is chosen for $a^*$ the rejection functions are given by:

$$R_l(a^*) = 1 - \Phi\left(\frac{a^* - \mu_l}{\sigma_l}\right) \tag{8.4a}$$

$$R_d(a^*) = \Phi\left(\frac{a^* - \mu_d}{\sigma_d}\right) \tag{8.4b}$$

where $\Phi(\cdot)$ is the standard normal cumulative distribution function; and $\mu_l$, $\mu_d$ and $\sigma_l$, $\sigma_d$ are the model's parameters. Other distributions like the log-normal (Equation 4.2) or Weibull (Equation 4.3) can also be chosen. To estimate $\mu$ _and_ $\sigma$ for the normal distribution, the likelihood functions are maximized by substituting $R_l(a^*)$ and $R_d(a^*)$ in Equations 8.3a and 8.3b by the expressions given in Equations 8.4a and 8.4b, respectively.

Figure 8.3 schematically shows a consumer who prefers a lighter-colored yogurt (A) and a consumer who prefers a darker-colored yogurt (B) and a dependency between the events "too light to JAR" and "JAR to too dark." If a consumer changes from too light to JAR in the region of very light colors, it is probable that he will change from JAR to too dark in the region of slightly dark colors (consumer (A) in Figure 8.3). Likewise, if a consumer changes from too light to JAR in the region of darker colors, it

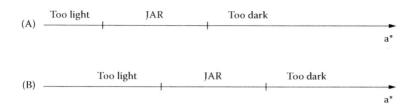

**Figure 8.3** Representation of a consumer who prefers a lighter-colored strawberry yogurt (A) and a consumer who prefers a darker-colored strawberry yogurt (B).

is probable that he will change from JAR to too dark in the region of very dark colors (consumer (B) in Figure 8.3).

This dependency will be analyzed with the following loglinear regression models with inclusion of covariates (Meeker and Escobar 1998; Garitta et al. 2006):

Light color rejection model:

$$\ln(a^*_1) = \mu_1 + \sigma_1\, W = \beta_{01} + \beta_{11}\, Y_1 + \sigma_1\, W \qquad (8.5a)$$

$a^*_1$ = color at which the consumer rejects the sample because it is too light

$\beta_{01}$ and $\beta_{11}$ = regression coefficients

$Y_1$ = covariate indicating if the consumer belongs to the category of those who reject lighter-colored samples because they are too dark ($Y_1 = 0$) or reject darker-colored samples because they are too light ($Y_1 = 1$)

$\sigma_1$ = shape parameter, which does not depend on the covariates

$W$ = error distribution

Dark color rejection model:

$$\ln(a^*_d) = \mu_d + \sigma_d\, W = \beta_{0d} + \beta_{1d}\, Y_d + \sigma_d\, W \qquad (8.5b),$$

$a^*_d$ = color at which the consumer rejects the sample because it is too dark

$\beta_{0d}$ and $\beta_{1d}$ = regression coefficients

$Y_d$ = covariate indicating if the consumer belongs to the category of those who reject lighter-colored samples because they are too dark ($Y_1 = 0$) or reject darker-colored samples because they are too light ($Y_d = 1$)

$\sigma_d$ = shape parameter, which does not depend on the covariates

$W$ = error distribution

As was mentioned in Section 5.3.1.2, when using category covariates such as $Y_l$ and $Y_d$ it is convenient to order the $Y$ values (0 and 1) in accordance with the alphabetical order of their names (dark and light). This simplifies the interpretation of the R-output.

### 8.2.2   Experimental data used to illustrate the methodology

A range of seven colors of liquid strawberry yogurt was used. The mid-range color was a leading commercial product. The lighter colors were obtained by mixing the yogurt with whole milk. The darkest colors were obtained by adding red coloring (Color 646, Christian Hansen, Quilmes, Buenos Aires, Argentina) to the commercial yogurt. The red color intensity of the samples, expressed as the *a\** CIELAB parameter, was measured with a Color-Tec PCM/PSMTM colorimeter (Color-Tec, Clinton, NJ). The a* values obtained were: 0.2, 3.9, 6.3, 10.9, 15.8, 21.4, and 25.3.

In Argentina, children are the age group who consume most liquid yogurt. Sixty children between 10 and 12 years, habitual consumers of liquid yogurt, were recruited. Each consumer received the seven samples corresponding to the seven colors monadically in random order. For each sample they had to mark one of the following options: "too light," "JAR," or "too dark." Further details on sample preparation, color measurement, and consumer study can be found in Garitta et al.'s (2006) article.

### 8.2.3   Optimum color calculations

To illustrate censoring, Table 8.2 presents response data from five of the consumers. As there are two events of interest, there are two censoring columns. Consumer 1 answers as expected, rejecting the very light

**Table 8.2** Acceptance and Rejection Data and Their Corresponding Censoring for Appearance of Liquid Yogurt Given by Five Consumers

| | Red color *a\** | | | | | | | Censoring | |
| | | | | | | | | Light color rejection | Dark color rejection |
| Consumer | 0.2 | 3.9 | 6.3 | 10.9 | 15.8 | 21.4 | 25.3 | | |
|---|---|---|---|---|---|---|---|---|---|
| 1 | L[a] | L | JAR | JAR | JAR | D | D | Interval: (3.9–6.3) | Interval: (15.8–21.4) |
| 2 | L | L | L | JAR | D | JAR | D | Interval: (6.3–10.9) | Interval: (10.9–25.3) |
| 3 | L | JAR | JAR | JAR | JAR | JAR | D | Left: < 3.9 | Interval: (21.4–25.3) |
| 4 | L | L | L | L | L | L | L | Right: > 25.3 | Right: > 25.3 |
| 5 | JAR | L | L | L | D | L | JAR | Not considered | |

[a] L: rejection due to too light, JAR: color is just about right, D: rejection due to too dark.

*Table 8.3* Censored Data for the Event *Too Light to JAR* in Yogurt Samples with Different Intensities of Red Color

| consumer | alow | ahigh | cens | censcod | dgroup | dmean |
|---|---|---|---|---|---|---|
| 1 | 3.89 | 6.32 | interval | 3 | dlight | 18.6 |
| 2 | 6.32 | 10.9 | interval | 3 | ddark | 23.345 |
| 3 | 10.9 | 15.83 | interval | 3 | ddark | 25.32 |
| ... | ... | ... | ... | ... | ... | ... |
| 58 | 10.9 | 15.83 | interval | 3 | dlight | 13.365 |
| 59 | 6.32 | 10.9 | interval | 3 | dlight | 18.6 |
| 60 | 10.9 | 15.83 | interval | 3 | ddark | 25.32 |

*Notes: alow* and *ahigh* are the lower and upper interval for each consumer; *dgroup* indicates whether the consumer belonged to the group who rejected samples with lower (*dlight*) or higher (*ddark*) values of red color because they were too dark; and *dmean* is average of the censored interval. The full data can be downloaded from the editor's Web site: light.xls.

samples, accepting those with intermediate colors, and rejecting very dark samples. Consumer 2 presents an inconsistency related to the "JAR to too dark" event, and this is reflected in a widening of his interval-censoring. Consumer 3 considered the first sample as too light, which is reasonable as it was 100% milk and thus completely white. This consumer found the second sample JAR; thus the data for this event was left-censored. It could be argued that the first sample did not have an a* value of 0, which would be the traditional condition for a left-censored observation; however, from a practical point of view, the first sample could be considered to have a red color = 0. Consumer 4 apparently likes his yogurt extremely red, because he even considered the last sample too light. Data for this consumer was right-censored for both observations. Consumer 5's data was not considered for two reasons:

*Found the first sample JAR.* This consumer could belong to a population who do not want their yogurts to be colored, but the search for an optimum color is based on the premise that targeted consumers want their yogurts colored. This reasoning is similar to what occurs when a consumer rejects the fresh sample in an SSL study (see Section 4.6).

*Highly inconsistent.* This can be clearly observed in Table 8.2. There is a tendency for this to occur with children and adolescents. The test situation distracts some of them to a point of total lack of concentration.

Tables 8.3 and 8.4 present the censored data for the too light to JAR and JAR to too dark events, respectively. This data can be obtained from the raw data shown in Table 8.2 by applying the R-function sslife (Figure 4.6 and Section 4.8):

1. To obtain Table 8.3 the D symbol (or whatever symbol has been used to indicate *too dark* or too much of the sensory variable being optimized) in Table 8.2 is changed to the JAR symbol. When using sslife the acceptance code should be L and the rejection code should be JAR. The *time* values would be: c(0.2, 3.9, 6.3, 10.9, 15.8, 21.4, and 25.3).

2. To obtain Table 8.4 the L symbol (or whatever symbol has been used to indicate *too light* or too little of the sensory variable being optimized) in Table 8.2 is changed to the JAR symbol. When using sslife the acceptance code should be JAR and the rejection code should be D. The *time* values would also be: c(0.2, 3.9, 6.3, 10.9, 15.8, 21.4, and 25.3).

3. The censored data that are displayed in R after using the sslife function can be copied and pasted to an Excel spreadsheet, one for the too light to JAR event and another for the JAR to too dark event.

4. The *dgroup* column of Table 8.3 indicates if the consumer belonged to the group who rejected samples with lower (*dlight*) or higher (*ddark*) values of red color because they were too dark. The *dmean* column is the average of the censored interval for this consumer corresponding to the JAR to too dark event. If *dmean* (expressed in $a^*$ values) was ≤ 18.6, *dgroup* = *dlight*; and if *dmean* > 18.6, *dgroup* = *ddark*.

5. The *lgroup* column of Table 8.4 indicates if the consumer belonged to the group who rejected samples with lower (*llight*) or higher (*ldark*) values of red color because they were too light. The *lmean* column is the average of the censored interval for this consumer corresponding to the too light to JAR event. If *lmean* (expressed in $a^*$ values) was ≤ 13.4, *lgroup* = *llight*; and if *lmean* > 13.4, *lgroup* = *ldark*.

**Table 8.4** Censored Data for the Event *JAR to Too Dark* in Yogurt Samples with Different Intensities of Red Color

| consumer | alow | ahigh | cens | censcod | lgroup | lmean |
|---|---|---|---|---|---|---|
| 1 | 15.83 | 21.37 | interval | 3 | llight | 5.105 |
| 2 | 21.37 | 25.32 | interval | 3 | llight | 8.61 |
| 3 | 25.32 | 25.32 | right | 0 | ldark | 13.365 |
| ... | ... | ... | ... | ... | ... | ... |
| 58 | 10.9 | 15.83 | interval | 3 | ldark | 13.365 |
| 59 | 15.83 | 21.37 | interval | 3 | llight | 8.61 |
| 60 | 25.32 | 25.32 | right | 0 | ldark | 13.365 |

*Notes:* *alow* and *ahigh* are the lower and upper interval for each consumer; *lgroup* indicates whether the consumer belonged to the group who rejected samples with lower (*llight*) or higher (*ldark*) values of red color because they were too light; and *lmean* is average of the censored interval. The full data can be downloaded from the editor's Web site: dark.xls.

The *dgroup* and *lgroup* variables were introduced to analyze possible dependency of one event on the other as expressed by Equations 8.5a and 8.5b. The limiting $a^*$ value of 18.6 was chosen to have approximately equal number of consumers classified as *dlight* and *ddark*; and, likewise, 13.4 was chosen to have approximately equal number of consumers classified as *llight* and *ldark*.

The detailed procedure to obtain the optimum color is as follows:

1. The data in the format shown in Tables 8.3 and 8.4 should be saved in tab-delimited format so that it can be read by R. Suppose that these files were named light.txt and dark.txt, respectively. These files must have the following columns:
   a. Column 1: consumer number. These numbers do not necessarily have to be consecutive nor start from 1; but all cells should be numbered and the column should include a text heading.
   b. Columns 2 and 3: upper and lower interval values of the designed parameter, in this case it is red color measured by $a^*$, corresponding to each consumer's censorship interval.
   c. Columns 4 and 5: type of censoring that corresponds to each of the consumers, represented by a text (Interval, Right, or Left) and a code (3, 0, and 2), respectively.
   d. Column 6: consumer classification factor. It is recommended that these groupings be represented by alphanumeric characters and not numbers.
2. The tab-delimited files are read into R:

```
> dark<- read.table("dark.txt",header=T)
> light<- read.table("light.txt",header=T)
```

3. Go to the File Menu, open Source R-code: sslcat.R (see Section 5.3.1.3 and Figure 5.4).
4. Use sslcat with the following options:

```
> resdark <- sslcat(dark,model="weibull")
> reslight <- sslcat(light,model="gaussian,"
percent=c(50,75,90))
```

The Weibull and normal (Gaussian) models presented the lowest values of loglikelihood for the full model, for the JAR to too dark and too light to JAR events, respectively. For the too light to JAR event, the rejection function is given by (1–distribution function), as for lower $a^*$ values rejection is higher than for higher $a^*$ values. Due to this, when using the sslcat R-function, if the $a^*$ value corresponding to a percent rejection = pr is to be calculated, the sslcat function has to be used for percent = c(100–pr). In the above application, for pr values of 50, 25, and 10%, percent = c(50,275,90).

5. Display the above results:

```
> resdark
```

```
$musig
```

| | parameter | estimate | lower | upper | znormal | prob100 |
|---|---|---|---|---|---|---|
| 1 | Intercept | 3.2716628 | 3.1326359 | 3.41068975 | 46.123858 | 0 |
| 2 | Llight | −0.1604703 | −0.3417041 | 0.02076354 | −1.735447 | 8.26616 |
| 3 | Sigma | 0.2595371 | 0.1902468 | 0.35406383 | NA | NA |

Considering Equation 8.5b:

$$\mu_d = \beta_{0d} + \beta_{1d} Y_d$$

```
$loglike
[1]  73.13669 71.57806
```

```
$chiprob100
[1]  7.74665
```

```
$slives
$slives$ldark
```

| percent | estimate | lower ci | upper ci | serror |
|---|---|---|---|---|
| 10 | 14.69649 | 11.93994 | 18.08945 | 1.557529 |
| 25 | 19.07361 | 16.31607 | 22.2972 | 1.519617 |
| 50 | 23.96367 | 20.89922 | 27.47747 | 1.6729 |

```
$slives$llight
```

| percent | estimate | lower ci | upper ci | serror |
|---|---|---|---|---|
| 10 | 12.51764 | 10.05968 | 15.57617 | 1.396122 |
| 25 | 16.24582 | 13.90257 | 18.98401 | 1.29106 |
| 50 | 20.41089 | 18.10063 | 23.01603 | 1.25092 |

The chi-square test on the significance of the full model given by Equation 8.5b was significant at a 7.7%, which showed a tendency. This same tendency can be observed in the subsequent tables, where, for example, for a 25% probability of rejecting a product because it was too dark, the $a^*$ values were 16 and 19, corresponding to consumers belonging to groups who had presented a transition from too light to JAR for lighter and darker samples, respectively. Thus the full model was adopted, and there will be two rejection curves for the JAR to too dark event, both responding to the Weibull model (see Equation 8.5b):

a. Group of consumers who changed from too light to JAR for $a^* > 13.4$

$$\mu_{dd} = \beta_{0d} + \beta_{1d} \, Y_d = 3.2717 - 0.1605 \times 0 = 3.2717$$

$$\sigma = 0.2595$$

b. Group of consumers who changed from too light to JAR for $a^* \leq 13.4$

$$\mu_{dl} = \beta_{0d} + \beta_{1d} \, Y_d = 3.2717 - 0.1605 \times 1 = 3.1112$$

$$\sigma = 0.2595$$

```
> reslight

$musig
```

| | parameter | estimate | lower | upper | znormal | prob100 |
|---|---|---|---|---|---|---|
| 1 | intercept | 11.503459 | 9.696218 | 13.310699 | 12.475802 | 0 |
| 2 | dlight | –3.229268 | –5.989587 | –0.46895 | –2.292984 | 2.184892 |
| 3 | sigma | 4.613843 | 3.701131 | 5.751633 | NA | NA |

```
$loglike
[1]  74.75598  72.23969

$chiprob100
[1]  2.487471

$slives
$slives$ddark
```

| percent | estimate | lower ci | upper ci | serror |
|---|---|---|---|---|
| 50 | 11.50346 | 9.696218 | 13.3107 | 0.9220617 |
| 75 | 14.61545 | 12.66855 | 16.56235 | 0.9933152 |
| 90 | 17.41634 | 15.165349 | 19.66732 | 1.1484624 |

`$slives$dlight`

| percent | estimate | lower ci | upper ci | serror |
|--------:|---------:|---------:|---------:|---------:|
| 50 | 8.27419 | 6.188086 | 10.36029 | 1.064339 |
| 75 | 11.38618 | 9.195808 | 13.57655 | 1.117536 |
| 90 | 14.18707 | 11.736823 | 16.63731 | 1.250125 |

The chi-square test on the significance of the full model given by Equation 8.5a was significant at a 2.49%. This difference can be observed in the subsequent tables, where, for example, for a 25% probability of rejecting a product because it was too light, the $a^*$ values were 11.4 and 14.6, corresponding to consumers belonging to groups who had presented a transition from JAR to too dark for lighter and darker samples, respectively. Thus, the full model was adopted, and there will be two rejection curves for the too light to JAR event, both responding to the normal distribution (see Equation 8.5a):

a. Group of consumers who changed from JAR to too dark for $a^* > 18.6$

$$\mu_{ld} = \beta_{01} + \beta_{11} Y_1 = 11.5035 - 3.2293 \times 0 = 11.5035$$

$$\sigma = 4.6138$$

b. Group of consumers who changed from JAR to too dark for $a^* \le 18.6$

$$\mu_{ll} = \beta_{01} + \beta_{11} Y_1 = 11.5035 - 3.2293 \times 1 = 8.2742$$

$$\sigma = 4.6138$$

6. Calculate the optimum:

From above there will be four different rejection curves: two corresponding to the too light to JAR event and two corresponding to the JAR to too dark event. These four curves are plotted in Figure 8.4. From these curves we can form two groups of consumers: those corresponding to curves (a) and (d) who prefer darker-colored yogurts; and those corresponding to (b) and (c) who prefer lighter-colored yogurts. These two consumer segments are illustrated in Figure 8.3. For each one of these consumer segments there will be an optimum color. The optimum is where the sum of percent rejection due to too light + percent rejection due to too dark is a minimum as illustrated in Figure 8.5 for the consumers who preferred lighter-colored yogurts. Numerically the optimum can be easily obtained by either of the following methods:

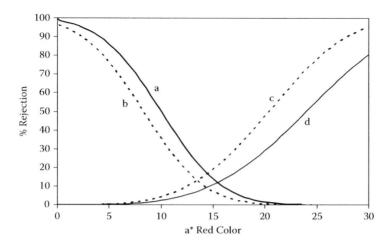

**Figure 8.4** Percent rejection versus red color of yogurts: (a) and (b): rejection due to *too light* for consumers who had presented a transition from JAR to too dark for darker and lighter samples, respectively; (c) and (d): rejection due to *too dark* for consumers who had presented a transition from too light to JAR for lighter and darker samples, respectively.

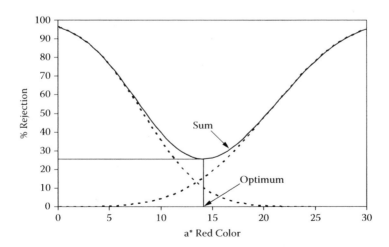

**Figure 8.5** Percent rejection versus red color of yogurts for consumers who preferred the lighter-colored products (curves (c) and (b) from Figure 8.4) and the sum of both rejection curves.

a. To plot Figure 8.5 four columns had to be formed in Excel: (1) for increasing $a^*$ values, (2) for the % rejection due to too light, (3) for the % rejection due to too dark, and (4) for the sum of (2) + (3). If column (1) has sufficiently small $a^*$ intervals, the $a^*$ value for which column (4) is minimum can be observed.

b. The Excel Solver tool can be applied to the four columns described in the previous paragraph.

The optimum colors for consumers' who preferred the lighter and darker yogurts were $a^* = 14.1$ and 17.5, respectively.

7. Obtain the optimum's confidence intervals.

As observed in Figure 8.5 at the optimum color value of 14.1 for consumers who preferred the lighter yogurts, there were consumers who would reject the sample because it was too light and consumers who would reject the sample because it was too dark. The total rejection probability was 25.6%, with 10.2% due to too light and 15.4% due to too dark. These last probabilities are taken as parameters of the sslcat R-function:

```
> resd <- sslcat(dark,model="weibull",
percent=c(15.4,50))
> resl <- sslcat(light,model="gaussian",
percent=c(50,89.8))
```

See Step 4 above for explanation as to why for the *light* data the percent values used in the sslcat function are introduced as 100-% rejection. The value of 50% is introduced, as a minimum of 2 values are needed for the sslcat function to work properly.

When resd$slives and resl$slives are displayed, the last column indicates the standard error of the estimations:

$llight

| percent | estimate | lower ci | upper ci | serror |
|---------|----------|----------|----------|---------|
| 15.4 | 14.11234 | 11.69387 | 17.03097 | 1.35352 |
| 50 | 20.41089 | 18.10063 | 23.01603 | 1.25092 |

$dlight

| percent | estimate | lower ci | upper ci | serror |
|---------|----------|----------|----------|---------|
| 50 | 8.27419 | 6.188086 | 10.36029 | 1.064339 |
| 89.8 | 14.13487 | 11.690639 | 16.57909 | 1.247055 |

The standard errors at the optimum for the JAR to too dark and too light to JAR events were 1.353 and 1.247, respectively. The confidence interval for an $\alpha$ significance level is:

$$optimal\ color \pm Z_{1-\alpha/2}se_{average} = optimal\ color \pm Z_{1-\alpha/2} \cdot \frac{1}{2} \cdot \sqrt{se_1{}^2 + se_2{}^2} \quad (8.6)$$

where

$Z_{1-\alpha/2}$ = $1-\alpha/2$ coordinate of the standard normal distribution

$se_1$ and $se_2$ = standard errors of the optimum color calculated from the rejection due to too light and rejection due to too dark curves, respectively

If Equation 8.6 is applied to the optimum corresponding to consumers who preferred the lighter products, the optimum ± 95% confidence intervals was $a^* = 14.12 \pm 1.80$. If the above calculations are applied to consumers who preferred the darker products, their optimum ± 95% confidence interval was $a^* = 17.5 \pm 1.82$.

## 8.2.4   Conclusions on optimum color estimations

For this product and the particular group of consumers, the optimum colors obtained by segmentation ($a^* = 14.1$ and $17.5$) were close to the color currently on the market at the time, which had an $a^*$ value of 15.8. The practical recommendation would be to keep the actual color on line.

As pointed out in previous chapters (Chapters 4 and 5), an important aspect of the survival analysis methodology is that experimental sensory work is relatively simple. In this work 60 consumers each looked at seven yogurt samples with different red color intensities, responding to whether they found the color too light, JAR, or too dark. This information was sufficient to determine an optimum color with an acceptable confidence interval and to segment consumers in groups according to whether they preferred lighter- or darker-colored yogurts. As discussed in Section 5.1, the recommended number of consumers is approximately 120. This estimation was not available at the time the optimum color methodology was developed; for future works 120 consumers are recommended.

In the local Argentine market there are already products targeted to segmented consumer groups, for example regular and hot ketchup, or regular and extra-strong mint sweets. The model proposed in the present work would be suitable in defining optimum pepper or mint concentrations for these segmented consumer groups.

## 8.3    Optimum salt level in French bread

In Argentina, consumption of French bread is estimated to be 190 grams per inhabitant, per day (SAGPYA 2005). Considering average sodium content in this type of bread, this represents an average daily sodium intake of 860 mg of sodium through bread (Sosa et al. 2008). The World Health Organization (WHO 2003) recommends a maximum daily salt intake of 2000 mg of sodium. Thus, through bread alone, Argentine population on average is covering 43% of the recommended daily salt intake. An important step in studying a possible salt reduction is to know what salt level people like.

Bread in Argentina is most often consumed accompanying a main meal; thus, a home-use test would seem more appropriate to measure its sensory acceptability when changing its level of saltiness. Sosa et al. (2008) considered it to be easier, from a logistic perspective, to have a consumer receive a single bread sample in her home instead of a number of samples on successive days, even if a larger number of consumers had to be contacted. In Section 5.2 the design where each consumer tests a single sample was presented.

In Sosa et al.'s (2008) work the influence of income on preferred level was considered. They hypothesized that low-income consumers, who generally have a lower education status, would be less aware of the dangers of high salt intakes and thus their preferred salt levels would be higher than those of middle-income consumers. Consumers' income status was considered as a covariate in Sosa et al.'s (2008) work; however, it was not significant. Bread is a staple food consumed equally by low-, middle-, and high-income populations in Argentina; thus this uniform behavior regarding preferred salt content. Another issue is that unless a person has been put on a strict low-salt diet, she does not consider staple foods such as bread to be a potential menace regarding salt intake.

Another issue of interest was the influence of the level normally consumed by the consumer on her preferred level. It can be hypothesized that if a consumer daily eats bread with a low salt content, her ideal salt level would be lower than a consumer who daily eats bread with a higher salt content. In the present illustration of current status data obtained from a home-use test, the sodium level normally consumed will be analyzed as a covariate.

### 8.3.1    Experimental data used to illustrate the methodology

Unless mentioned otherwise sodium concentration units shall all be understood as milligrams of sodium per 100 grams of bread on a dry basis. Sosa et al. (2008) conducted an initial experiment to determine the

*Table 8.5* Percent Sodium Chloride and Sodium
Concentration in Resulting Bread

| % NaCl (g/100 g flour) | Sodium (mg Na/100 g dry bread) |
|---|---|
| 0.6 | 272 |
| 1.2 | 532 |
| 1.8 | 789 |
| 2.4 | 1043 |
| 3.0 | 1293 |
| 3.6 | 1540 |
| 4.2 | 1785 |

concentration difference in bread that produced a barely noticeable difference in saltiness. They obtained a value of 176 mg Na; from this just noticeable difference a step of approximately 260 mg Na was adopted for preparation of bread samples with different salty taste to make sure there was a perceivable difference among samples. Small variations were made in this step to have a regular 0.6 grams NaCl/100 grams flour step in the bread formulation. Final concentrations are in Table 8.5. Yeast and enzyme additive values varied with the salt content to avoid major texture changes due to salt variations. The bread samples were manufactured in a local bakery shop following their standard procedure for French bread. Consumers tasted the bread samples 2 to 4 hours after baking.

Each sample was received by 60 consumers, and each consumer received only one salt concentration. As there were seven sodium concentrations (see Table 8.5), so this meant a total of 420 consumers. In Section 5.1 the recommended number of consumers necessary for shelf-life estimations based on survival analysis statistics was 120. There are no published estimates of necessary number of consumers where each consumer receives a single sample. Araneda et al. (2007) performed a study on shelf life of lettuce where each consumer evaluated a single sample, and they used 50 consumers per sample and obtained estimations with adequate 95% confidence intervals. The bread samples were delivered in brown paper bags, and instructions were given to the adults receiving the samples to consume at least one loaf (approximately 100 grams) during lunch. After eating the bread they had to mark one of the following options: *not salty enough*, *JAR*, or *too salty*. They were also asked where they normally bought their bread. Their answer sheets were collected during the afternoon of the same day they had received the sample.

In order to establish a relationship between preferred salty taste and the salt content eaten daily by consumers, bread samples were taken from the bakery shops where consumers normally bought their bread (a total of 27 shops) and their sodium and moisture contents were analyzed.

## 8.3.2   Survival analysis model

The discussion on censoring for both events of interest was presented above (Section 8.2.3) for a study where each consumer tasted all concentrations. For a home use test where each consumer tasted only one concentration, the censoring scheme is similar to that presented for the cracker current-status study presented in Section 5.2. In the cracker study there was only one event of interest: from JAR to deteriorated. In the present case, with two events of interest, the censoring scheme is exemplified as follows:

- If a consumer evaluated bread with a concentration of 789 mg Na and found it not salty enough, the event *not salty enough to JAR* was right-censored, and the event *JAR to too salty* was also right-censored.
- If a consumer evaluated bread with a concentration of 789 mg Na and found it JAR, the event *not salty enough to JAR* was left-censored, and the event *JAR to too salty* was right-censored.
- If a consumer evaluated bread with a concentration of 789 mg Na and found it too salty, the event *not salty enough to JAR* was left-censored, and the event *JAR to too salty* was also left-censored.

Salt level in bread consumed daily (low, middle, and high) was introduced as a covariable in the analysis of results. Each category contained the following number of consumers:

- Low (less than 550 mg Na): 86
- Medium (between 550 and 650 mg Na): 203
- High (more than 650 mg Na): 131

The high number of consumers in the medium category was due to 94 consumers who answered they bought their bread "anywhere" and were assigned the mean concentration over the 27 bakeries. The concentration limits to classify consumers were chosen based on round numbers and to have approximately equal number of consumers in each category, without counting those who answered "anywhere."

The influence of the covariate on the rejection salt concentrations was analyzed with the following loglinear regression models with inclusion of covariates (Meeker and Escobar 1998):

*Not salty enough* rejection model:

$$\ln(c_{nse}) = \mu_{nse} + \sigma_{nse}\, W = \beta_{0nse} + \beta_{1nse}\, Y_{1nse} + \sigma_{nse}\, W \tag{8.7a}$$

where

$c_{nse}$ = concentration at which the consumer rejects the sample because it is not salty enough

$\beta_{0nse}, \beta_{1nse}$ = regression coefficients

$Y_{1nse}$ = covariate indicating whether the consumer normally consumed bread with low, medium, or high salt content

$\sigma_{nse}$ = shape parameter, which does not depend on the covariates

$W$ = error distribution

Analogously, the *too salty* rejection model was:

$$\ln(c_{ts}) = \mu_{ts} + \sigma_{ts} \, W = \beta_{0ts} + \beta_{1ts} \, Y_{1ts} + \sigma_{ts} \, W \qquad (8.7b)$$

Once the likelihood is formed for a given model, specialized software can be used to estimate the parameters ($\beta$ coefficients and $\sigma$) that maximize the likelihood function for the given experimental data.

## 8.3.3   Optimum salt concentration calculations

Tables 8.6 and 8.7 present data for 8 of the 420 consumers in the form to be read by R for the not salty enough to JAR and for the JAR to too salty events, respectively. Calculations of the optimum salt concentrations are

*Table 8.6* Censored Data for Six Consumers for the Event *Not Salty Enough to JAR* in Bread Samples with Different Salt Concentrations

| consumer | saltl | salth | cens | censcod | bakery |
|----------|-------|-------|------|---------|--------|
| 1 | 272 | 272 | right | 0 | bmed |
| 2 | 272 | 272 | right | 0 | bmed |
| 4 | 272 | 272 | right | 0 | bmed |
| ... | ... | ... | ... | ... | ... |
| 218 | 1043 | 1043 | left | 2 | chigh |
| 220 | 1043 | 1043 | right | 0 | alow |
| ... | ... | ... | ... | ... | ... |
| 418 | 1785 | 1785 | left | 2 | chigh |
| 419 | 1785 | 1785 | left | 2 | bmed |
| 420 | 1785 | 1785 | left | 2 | bmed |

*Notes:* *saltl* and *salth* are the lower and upper intervals for each consumer; *bakery* indicates whether the consumer belonged to the group who normally consumed bread with low (alow), medium (bmed), or high (chigh) salt concentrations. The full data can be downloaded from the editor's Web site: breadnse.xls.

*Table 8.7* Censored Data for Six Consumers for the Event *JAR to Too Salty* in Bread Samples with Different Salt Concentrations

| consumer | saltl | salth | cens | censcod | bakery |
|---|---|---|---|---|---|
| 1 | 272 | 272 | right | 0 | bmed |
| 2 | 272 | 272 | right | 0 | bmed |
| 4 | 272 | 272 | right | 0 | bmed |
| ... | ... | ... | ... | ... | ... |
| 218 | 1043 | 1043 | right | 0 | chigh |
| 220 | 1043 | 1043 | right | 0 | alow |
| ... | ... | ... | ... | ... | ... |
| 418 | 1785 | 1785 | left | 2 | chigh |
| 419 | 1785 | 1785 | left | 2 | bmed |
| 420 | 1785 | 1785 | left | 2 | bmed |

*Notes: saltl* and *salth* are the lower and upper intervals for each consumer; *bakery* indicates whether the consumer belonged to the group who normally consumed bread with low (alow), medium (bmed), or high (chigh) salt concentrations. The full data can be downloaded from the editor's Web site: breadtoo.xls.

similar to those detailed for calculating the optimum color in Section 8.2.3; thus, some details will not be repeated. The steps are as follows:

1. The tab-delimited files are read into R:

```
> nse<- read.table("breadnse.txt",header=T)
> too<- read.table("breadtoo.txt",header=T)
```

2. Go to the File Menu, open Source R-code: sslcat.R (see Section 5.3.1.3 and Figure 5.4).
3. Use sslcat with the following options:

```
> resnse <- sslcat(nse,model="weibull",
percent=c(50,75,90))
> restoo <- sslcat(too,model="weibull")
```

The loglikelihood values for the Weibull, log-normal, and Gaussian models were similar; the Weibull was chosen as it is widely used in survival and reliability studies. For the not salty enough to JAR event, the rejection function is given by (1–distribution function); as for lower salt concentrations, rejection is higher than for higher concentrations. Due to this, when using the sslcat R-function, if the concentration value corresponding to a percent rejection = pr is to be calculated, the sslcat function has to be used for percent = c(100–pr).

In the above application, for pr values of 50, 25, and 10%, percent = c(50,275,90).

4. Display the chi-square probability values:

```
> resnse$chiprob100
[1]  68.35326
> restoo$chiprob100
[1]  27.33853
```

It is clear that these chi-square probabilities are not significant ($P > 5\%$); thus the inclusion of the bakery covariable was not significant.

5. As the covariate was not significant, the model's parameters are recalculated without the inclusion of the covariate. For this, go to the File Menu, open Source R-code: sslcsd.R (see Section 5.2.3 and Figure 5.2).

6. Use sslcsd with the following options:

```
> resnse <- sslcsd(nse,model="weibull",
percent=c(50,75,90))
> restoo <- sslcsd(too,model="weibull")
```

The loglikelihood values for the Weibull, log-normal, and Gaussian models were similar; the Weibull was chosen as it is widely used in survival and reliability studies.

7. The parameters of the Weibull model for each event were the following:
   - Not salty enough to JAR: $\mu = 6.41$ and $\sigma = 0.758$
   - JAR to too salty: $\mu = 7.40$ and $\sigma = 0.205$

Figure 8.6 shows the rejection curves corresponding to not salty enough and too salty, and the sum of both curves whose minimum corresponds to the optimum sodium concentration. This optimum was 985 mg, for which there was a 23.6% rejection probability; 15.5% and 8.1% corresponded to not salty enough and too salty, respectively. These last probabilities are taken as parameters of the sslcsd R-function:

```
> resnse <- sslcsd(nse,model="weibull",
percent=c(50,84.88))
> restoo <- sslcsd(too,model="gaussian",
percent=c(8.1,50))
```

When resnse$slives and restoo$slives are displayed, the last column indicates the standard error of the estimations:

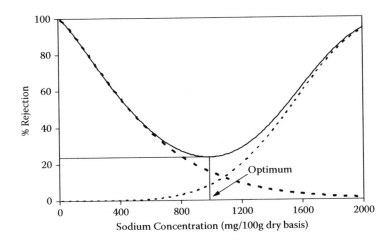

***Figure 8.6*** Percent rejection versus salt concentration in bread for *not salty enough* and for *too salty*, and the sum of both rejection curves. The optimum is the minimum of this last curve. (Reprinted with permission from: Sosa, M., Flores, A., Hough, G., Apro, N., Ferreyra, V., and Orbea, M.M. 2008. Optimum level of salt in French-type bread: influence of income status, salt level in daily bread consumption and test location. *Journal of Food Science* 73: S392–397.)

```
> resnse$slives
```

| percent | estimate | lower ci | upper ci | serror |
|---|---|---|---|---|
| 50 | 460.6438 | 388.7531 | 545.8289 | 39.8787 |
| 84.88 | 984.7914 | 884.2654 | 1096.7455 | 54.0995 |

```
> restoo$slives
```

| percent | estimate | lower ci | upper ci | serror |
|---|---|---|---|---|
| 8.062 | 984.9644 | 887.4662 | 1093.174 | 52.38152 |
| 50 | 1518.1048 | 1456.5814 | 1582.227 | 32.04328 |

The standard errors at the optimum for the not salty enough and too salty events were = 54.1 and 52.4, respectively. Applying Equation 8.6 the optimum ± 95% confidence interval was 985 ± 74 mg sodium.

### 8.3.4   Conclusions on optimum salt concentration estimation

Sosa et al. (2008) discussed the hypothesis that consumers who daily consumed bread with a low salt concentration would have their preferred salt levels shifted to lower concentrations; and inversely for consumers who daily consumed bread with a high salt concentration. This was not the case, as salt content of bread consumed daily was neither significant in rejection probability as not salty enough, nor in rejection probability as too salty. In principle this non-significance would seem unlikely. The average salt content expressed as sodium in bread from bakeries where consumers bought their daily bread was 628 mg Na with minimum and maximum values of 438 mg Na and 804 mg Na, respectively. The estimated optimum salt level was 980 mg Na with only 8% probability of consumers rejecting this level for being too salty. Thus, most consumers, regardless of where they buy their daily bread, would really like their bread saltier. If a main entrée of a meal consists of roasted chicken and mashed potatoes, a consumer, out of habit, will judge the saltiness of each component of the entrée. He may decide the chicken is okay but the mashed potatoes are not salty enough and ask for the salt shaker. But also out of habit there won't be a judgment on the saltiness of the accompanying bread. It is very rare to put salt on bread. Thus, consumers in their daily consumption do not register the saltiness of bread, but when their attention is focused on this issue, as it was in this work, they prefer higher saltiness than what they are used to.

Survival analysis methodology proved an adequate tool to obtain an estimation of the optimum concentration of a food ingredient where each consumer evaluated a single sample. This methodology was particularly adequate for a product like bread, which is normally eaten as an accompanying staple food, and thus a home use test was called for.

## 8.4   Internal cooking temperature of beef

Color changes are perceived by consumers when evaluating the degree of meat doneness. In the United States, the Beef Steak Color Guide (American Meat Science Association, Chicago) illustrates these color changes from *very rare* to *well-well done* beef. This color guide has six pictures of beef steaks labeled as 55°C—very rare, 60°C—rare, 63°C—medium rare, 71°C—medium, 77°C—well done, and 82°C—well well done. López Osornio et al. (2008) applied survival analysis statistics models to the prediction of optimum cooking temperatures of beef based on acceptance or rejection data obtained from consumers. Consumers from different countries, age groups, and stated preferences for degree of doneness were considered.

Consumers who consumed cooked beef at least once a week were recruited from Nueve de Julio (Argentina), Manhattan, Kansas (United States), and Valencia (Spain). In each location 102 consumers were recruited, half aged 21 to 30 years (young) and half aged 40 to 60 years (middle-aged). Color guides were acquired from the National Cattlemen's Beef Association (Centennial, Colorado (United States)).

The pictures were cut out and coded with three-digit random numbers. Consumers received the six pictures monadically in balanced order. For each picture consumers had to tick a box indicating whether they considered the meat in the picture *undercooked*, *JAR*, or *overcooked*. After evaluating the six pictures, consumers were asked how they normally consumed their beef; their answers were categorized as rare, medium, and well-done.

The survival analysis model was basically the same as the one described above for determining the optimum color of yogurt (Section 8.2.1). In that model the explanatory variable was the color of the yogurt expressed by the Hunter Lab $a^*$ parameter. In the present case, the explanatory variable was internal cooking temperature (ICT).

Let $T$ be the random variable representing the optimum ICT of beef for a given consumer. Assume that $T$ is absolutely continuous with distribution function $F$. For each value of ICT $t$, there will be two rejection functions:

$R_u$ $(t)$ = probability of a consumer (or proportion of consumers) rejecting beef with ICT = $t$ because it is undercooked; that is, $R_u(t) = P(T > t) = 1 - F(t)$

$R_o$ $(t)$ = probability of a consumer (or proportion of consumers) rejecting beef with ICT = $t$ because it is overcooked, that is $R_o$ $(t) = P$ $(T < t) = F$ $(t)$

The undercooked rejection model is expressed by:

$$\ln(t_u) = \mu_u + \sigma_u W = \beta_{0u} + \beta_{1u} Z_{1u} + \beta_{2u} Z_{2u} + \beta_{3u} Z_{3u} +$$
$$[\text{two-way interactions}] + \sigma_u W \tag{8.8}$$

where

$t_u$ is the ICT at which a consumer rejects a sample because it is undercooked

$\beta_{0u}$, $\beta_{1u}$, $\beta_{2u}$ and $\beta_{3u}$ are the regression coefficients

$Z_{1u}$ is the covariate indicating the country of residence: Argentina, Spain, or United States

$Z_{2u}$ is the covariate indicating stated preference for degree of doneness: medium, rare, or well-done

$Z_{3u}$ is the covariate indicating consumer age group: middle-aged or young

$\sigma_u$ is the shape parameter, which does not depend on the covariates

$W$ is the error distribution

An analogous expression to Equation 8.8 was constructed for the overcooked rejection model.

The sslcat R-function, which was used in Chapter 5 (Section 5.3.1.3) and above (Sections 8.2.3 and 8.3.3), can only deal with a single covariate. For the present model with more than one covariate, the more basic survreg R-function associated with the predict R-function, can be used. A more friendly, albeit more costly, alternative is to use the Survival-Life testing menu of TIBCO Spotfire S+ (TIBCO Inc., Seattle, WA) software.

A loglikelihood test (Meeker and Escobar 1998) indicated no significant differences between the models with or without the two-way interaction terms. This was valid both for the undercooked event and the overcooked event.

Consumers' country of origin effect was significant, although the magnitude of differences was small. Argentine consumers tended to prefer higher ICTs, followed by Spanish and U.S. consumers. The age effect was significant but its magnitude was small; young consumers tended to prefer slightly lower ICTs than older consumers.

The factor that most influenced both undercooked and overcooked rejection was stated preference for degree of doneness. Figure 8.7 shows the optimum cooking temperatures for young Argentine consumers (sum of undercooked rejection + overcooked rejection) who preferred rare, medium, and well-done beef, these optimums were 75°C, 78°C, and 82°C, respectively.

One of the conclusions of López Osornio et al.'s (2008) work was that optimum temperatures were relatively high, and that only 4% of the

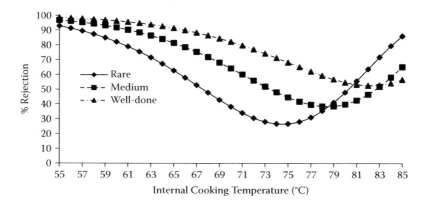

*Figure 8.7* Optimum cooking temperatures for young Argentine consumers (sum of undercooked rejection + overcooked rejection) whose stated degree of preference for beef was rare, medium, and well done. (Reprinted with permission from: López Osornio, M.M., Hough, G., Salvador, A., Chambers IV, E., McGraw, S., and Fiszman, S. 2008. Beef's optimum internal cooking temperature as seen by consumers from different countries using survival analysis statistics. *Food Quality and Preference* 19: 12–20.)

consumers found the 55°C—very rare and 60°C—rare pictures JAR. This suggested the necessity of revising the range of the AMSA Color Guide (American Meat Science Association, Chicago).

Survival analysis methodology was an adequate tool to estimate an optimum ICT of beef, based on consumer acceptability. As in previously described experiments the sensory work was relatively simple; only 102 consumers had to decide whether they found successive samples undercooked, JAR, or overcooked. This methodology could be extended to determine optimum cooking times and/or temperatures of other food products.

## 8.5   Optimum ripening times of fruits

In Section 8.2.1 the model published by Garitta et al. (2006) to estimate optimum concentrations of a food ingredient using survival analysis statistics was presented. The model was applied to red color in a strawberry yogurt, where the color can be too light, JAR, or too dark, leading to two events of interest: the transition of too light to JAR, and the transition of JAR to too dark. This is a similar scenario to fruit ripening, where the fruit can be underripe, okay, or overripe. Garitta el al. (2008) applied survival analysis statistics models to the prediction of optimum ripening stages of tomatoes based on acceptance or rejection data obtained from consumers.

In the case of the yogurt color discussed in Section 8.2, consumers were segmented into those who preferred lighter-colored yogurts and those who preferred darker-colored yogurts, as shown in Figure 8.3. Similarly, when estimating the parameters for the overripe event in tomatoes it is possible that the rejection as overripe is dependent on the rejection as underripe. For example, if a consumer rejects only the very green tomatoes because they are underripe, she will probably start rejecting only slightly ripe tomatoes because they are overripe; that is, her accepted ripeness times are low. Analogously, if a consumer rejects relatively ripe tomatoes because they are underripe, she will probably start rejecting very ripe tomatoes because they are overripe; that is, her accepted ripeness times are high. If we assume that consumers have a previous background on tomato ripeness choice, then the rejection due to underripe could also be dependent on what tomato they will consider to be overripe, in spite of the fact that in Garitta et al's (2008) experiment consumers received the tomatoes in ascending ripeness order. These considerations lead to underripe and overripe rejection models expressed by equations similar to Equations 8.5a and 8.5b, respectively:

Underripe rejection model:

$$\ln(t_u) = \mu_u + \sigma_u W = \beta_{0u} + \beta_{1u} Y_u + \sigma_u W \qquad (8.9a)$$

where

$t_u$                    = time at which the consumer rejects the sample because it
                         is underripe

$\beta_{0u}$ and $\beta_{1u}$ = regression coefficients

$Y_u$                   = covariate indicating whether the consumer belongs to the
                         category of those who reject low ripeness-time tomatoes
                         because they are overripe or reject higher ripeness-time
                         tomatoes because they are overripe

$\sigma_u$               = shape parameter, which does not depend on the
                         covariates

$W$                     = error distribution

Overripe rejection model:

$$\ln(t_o) = \mu_o + \sigma_o W = \beta_{0o} + \beta_{1o} Y_o + \sigma_o W \qquad (8.9b)$$

where

$t_o$                    = time at which the consumer rejects the sample because it is
                         overripe

$\beta_{0o}$ and $\beta_{1o}$ = regression coefficients

$Y_o$                   = covariate indicating whether the consumer belongs to
                         the category of those who reject low ripeness-time toma-
                         toes because they are underripe or reject higher ripeness-
                         time tomatoes because they are underripe

$\sigma_o$               = shape parameter, which does not depend on the
                         covariates

$W$                     = error distribution

Tomatoes (*Lycopersicum esculentum* var. Coloso) were obtained from a local producer (Nueve de Julio, Buenos Aires, Argentina). Day 0 corresponded to completely green tomatoes except for first signs of red color at the base. Tomatoes were collected daily till day 7. Details of the collecting procedure are in Garitta et al.'s (2008) paper.

Sixty regular consumers of fresh tomatoes who were willing to perform daily evaluations for the duration of the study were recruited. For appearance all consumers evaluated the same two tomatoes placed on a white plate under artificial daylight-type illumination. For flavor they received a slice of tomato in a plastic dish. For both appearance and flavor they had to answer whether they found the sample underripe, okay, or overripe.

Time = 0 corresponded to completely green tomatoes except for first signs of red color at the base. It was supposed that at this condition all consumers would reject the tomato for being underripe, and this is what occurred in Garitta et al.'s (2008) study. However if in a fruit ripeness study more than 10% of the consumers accept the fruit at time = 0, it is probable that this initial time was too late and the fruit was slightly ripe, leading to acceptance by some consumers. The data for these consumers would be left-censored; that is, the event *underripe to okay* occurred before the first registered time. If this were the case, time = 0 would have to be fixed at some previous time by consulting an agronomist who would determine a time when the fruit was definitely too green.

For appearance consumers were not segmented; that is, the covariates of Equations 8.9a and 8.9b were not significant. This means that consumers were uniform in their rejection curves for underripeness and overripeness of tomatoes. For flavor there was partial segmentation. The covariate in Equation 8.9a was significant; that is, a consumer's rejection due to underripe tended to be influenced by when they considered a tomato to be overripe. The covariate in Equation 8.9b was not significant for flavor.

As the covariates were not significant for appearance, there was a single optimum ripening time for this attribute; the value ± 95% confidence interval was 3.4 ± 0.7 days. For flavor there were two optimum values corresponding to the two consumer segments defined by the covariate in Equation 8.9a that was significant:

1. For consumers who rejected the flavor of tomatoes due to overripeness with ripening times ≤6.5 days: 2.7 ± 0.9 days
2. For consumers who rejected the flavor of tomatoes due to overripeness with ripening times >6.5 days: 3.5 ± 0.8 days

Figure 8.8a and 8.8b show rejection curves for appearance and flavor (segment 2 above), respectively; together with their corresponding optimum ripening times. The following observations can be made on these figures:

- At the optimum ripening times the percent rejections were relatively low: 4.3% for appearance and 7.1% for flavor. These values were lower than those found in the studies for color in yogurt, salt in bread, and temperature in beef, where percent rejections at the optimums were >20% (Figures 8.5, 8.6, and 8.7). This shows that for tomato-ripening times there was a high uniformity among the consumers. This uniformity could have been a consequence of the experimental design. In Section 3.3.6.1 the basic design was presented for shelf-life studies. It consists of storing a single batch at the desired temperature and periodically removing samples from storage and analyzing them (see Figure 3.3). The design used for the ripening time of tomatoes was similar to the basic design. A drawback of this design is that consumers can become aware that they are participating in a

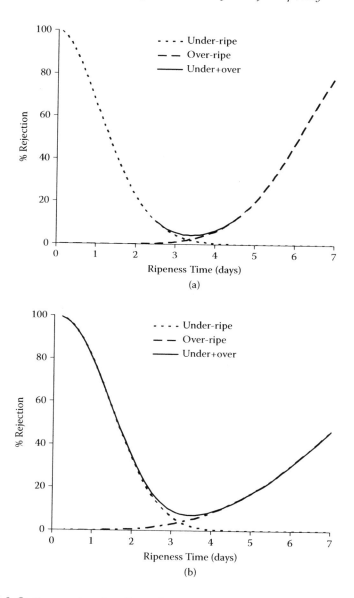

*Figure 8.8* Optimum ripening times for tomatoes: (a) appearance and (b) flavor. (Reprinted with permission from: Garitta, L., Hough, G., and Hulshof, E. 2008. Determining optimum ripening time of fruits applying survival analysis statistics to consumer data. *Food Quality and Preference* 19: 747–752.)

sensory shelf-life study or, as in this case, in a fruit-ripening study. This awareness could have led to biased results. When leaving or arriving for the different sessions, consumers could have commented on what they were evaluating, thus tending to uniform responses.

- The optimum ripening time for appearance and flavor was similar, but as ripening progressed tomato appearance was more important than flavor in the rejection of the fruit. The overripe rejection curve for appearance had a higher slope than the overripe rejection curve for flavor. This led to increasing rejection probability for appearance in relation to flavor. For 7 days' ripening time there were 78% and 46% rejections for appearance and flavor, respectively.

As in the previous cases presented, experimental sensory work applied in this study was relatively simple; only 60 consumers available during the ripening period had to decide whether they found the successive samples underripe, okay, or overripe. This censored data set was sufficient to determine the optimum ripeness time and to segment consumers who liked the fruits in the underripened range and consumers who liked the fruits in the overripened range.

## References

Araneda, M., G. Hough, and E. Wittig de Penna. 2008. Current-status survival analysis methodology applied to estimating sensory shelf life of ready-to-eat lettuce (lactuta sativa). *Journal of Sensory Studies* 23: 162–170.

Conner M.T., A.V. Haddon, and D.A. Booth. 1986. Very rapid, precise assessment of effects of constituent variation on product acceptability: Consumer sweetness preferences in a lime drink. *Lebensmittel Wissenschraft und Technologie* 19: 486–490.

Costell, E. 2002. A comparison of sensory methods in quality control. *Food Quality and Preference* 13: 341–353.

Garitta, L., G. Hough, and E. Hulshof. 2008. Determining optimum ripening time of fruits applying survival analysis statistics to consumer data. *Food Quality and Preference* 19: 747–752.

Garitta, L., C. Serrat, G. Hough, and A. Curia. 2006. Determination of optimum concentrations of a food ingredient using survival analysis statistics. *Journal of Food Science* 71: S526–532.

Hernandez, S.V., and H.T. Lawless. 1999. A method for adjustment for preferred levels of capsaicin in liquid and solid food systems among panelists of two ethnic groups and comparison to hedonic scaling. *Food Quality and Preference* 10: 41–49.

Hough, G., L. Garitta, and R. Sánchez. 2004. Determination of consumer acceptance limits to sensory defects using survival analysis. *Food Quality and Preference* 15: 729–734.

Hutchings, J.B. 1994. *Food color and appearance.* New York: Chapman & Hall.

Klein, J.P., and M.L. Moeschberger. 1997. *Survival analysis, techniques for censored and truncated data.* New York: Springer-Verlag Inc.

Lawless, H.T., and H. Heymann. 1998. *Sensory evaluation of food, principles and practices.* New York: Chapman & Hall.

López Osornio, M.M., G. Hough, A. Salvador, E. Chambers IV, S. McGraw, and S. Fiszman. 2008. Beef's optimum internal cooking temperature as seen by consumers from different countries using survival analysis statistics. *Food Quality and Preference* 19: 12–20.

Meeker, W.Q., and L.A. Escobar. 1998. *Statistical methods for reliability data.* New York: John Wiley & Sons.

Muñoz, A.M. 2002. Sensory evaluation in quality control: An overview, new developments and future opportunities. *Food Quality and Preference* 13: 329–339.

Muñoz A., G.V. Civille, and B.T. Carr. 1992. *Sensory evaluation in quality control,* pp. 80–81. New York: Van Nostrand Reinhold.

Rothman, L., and M.J. Parker. 2009. *Just-About-Right (Jar) scales: design, usage, benefits and risks,* pp. 1–13, ASTM Number MNL63. West Conshohocken, PA: American Society for the Testing and Materials (ASTM).

SAGPYA. 2005. Secretaría de Agricultura, Ganadería, Pesca y Alimentación. Available from: http://www.alimentosargentinos.gov.ar/0-3/farina/Panificados/ptos panificados 12 05.htm. Accessed Nov. 8, 2008.

Sosa, M., A. Flores, G. Hough, N. Apro, V. Ferreyra, and M.M. Orbea. 2008. Optimum level of salt in French-type bread. Influence of income status, salt level in daily bread consumption and test location. *Journal of Food Science* 73: S392–397.

Weller, J.N., and K.J. Stanton. 2002. The establishment and use of a QC analytical/descriptive/consumer measurement model for the routine evaluation of products at manufacturing facilities. *Food Quality and Preference* 13: 375–383.

WHO. 2003. Diet, nutrition and the prevention of chronic diseases. WHO Technical Report Series 916. Geneva, Switzerland: WHO.

# *Index*